Changing Roles of Agricultural Extension in Asian Nations

Changing Roles of Agricultural Extension in Asian Nations

Editors

A.W. Van den Ban and R.K. Samanta

B. R. Publishing Corporation
[A Division of BRPC (India) Ltd.]
Delhi-110052

Changing Roles of Agricultural Extension in Asian Nations

A.W. Van den Ban and R.K. Samanta

First Published 2006

ISBN 81-7646-573-9

Rs. 650/-

Published by :

B. R. Publishing Corporation
[A Division of BRPC (India) Ltd.]
425, Nimri Colony,
Ashok Vihar, Phase-IV
Delhi – 110052
E-Mail : *brpcltd@del2.vsnl.net.in*

Printed In India at Balaji Offset New Delhi.

Preface

In many Asian countries, there is now a more rapid economic growth than there has ever been in the so called developed countries. This results in major changes in agriculture and therefore agricultural extension also has to change. Policy makers have to decide which changes in extension are desirable in their country and how these changes can be realized. This implies learning from experience. Valuable information can not only be obtained from the experience in their own country, but also from other countries, where the society is changing in a somewhat similar way. However, it is often difficult to obtain information about these changes in extension in many Asian countries.

That was for us a reason to invite people from universities, research institutes, NGOs and other extension and donor organizations to write a chapter for this book about the changes in the roles of agricultural extension in which they have been involved. They come from eight Asian countries and several of them have worked in many other countries. We are grateful for their contributions to this book which documents their experience with the changing role of agricultural extension and we hope that Asian farmers will benefit from their efforts. Despite the economic growth many of these farmers are still quite poor and a major role of agricultural extension is to try to change this.

Farmers, and certainly poor farmers, can only profit from the opportunities the economic growth in their country offers, if their entrepreneurial and managerial capabilities increase. This book tries to find ways to realize this. Further, the present book is expected to be of greater help to policy planners, agricultural development administrators, NGOs, researchers, extension teachers and students in their endeavours to contribute in the development process throughout Asian nations.

We also take this opportunity to express our sincere thanks and appreciation to Mr. Praveen Mittal, Prop. of B.R.Publishing Corporation, India Ltd. for ensuring prompt production and publication of the book, and Mrs. K.K. Rukmani Ammal for neatly type setting the book's manuscript.

Wageningen, The Netherlands and
Hyderabad, India
March 2006

A.W. Van den Ban
R.K. Samanta

Contents

1

Introduction Agricultural Extension in Asian Nations : Why it needs Change!

A.W. van den Ban and R.K. Samanta

Prologue

In this chapter we discuss briefly changes in the roles of agricultural extension we observed in Asian nations or which might be considered desirable there and the changes in Asian societies which make it necessary to change agricultural extension. Many of these changes are discussed in more detail in this book by chapter authors who have been directly involved in these processes of change. We hope that these discussions will help readers to consider which changes are desirable in their situation and which are not and how the desirable changes can be implemented. Clearly there are large differences in socio-economic situation between different Asian Nations and even between regions within one nation. Therefore changes which are successful in one nation, may not be desirable or possible in another nation or region.

Changes in Society which cause Opportunities and Threats for Farmers

Economic growth

During the last decade we have seen a more rapid economic growth in many Asian Nations than we have ever seen in Europe or America. This has caused an increase in demand for high value agricultural products, labour costs and employment opportunities outside agriculture. Farmers have to adjust the management of their enterprise to this changed economic environment in order to be able to compete in the market.

Increased agricultural productivity

The Green revolution has resulted in a large increase in the yields of cereals and therefore in a major decrease in the number of people suffering from hunger. This does not mean that hunger has been abolished. In large parts of South Asia still more than 40% of the pre-school children are underweight (Sanchez and Swaminathan, 2005: 51), but this is more because their family is too poor to buy enough food for children than the non-availability of food in the market.

The increase in yields has resulted in an increase in agricultural production and therefore in a decrease in prices. Between 1980 and 2000 the world market prices for agricultural commodities decreased with about 50% (FAO, 2002: 12, World Bank, 2005: 23). This means that farmers have to increase their productivity a lot, if they like to maintain their level of income. If they like to increase their income as much as urban people in their country an even larger increase in productivity is necessary. We do not know what will happen with the prices of agricultural commodities in the next decades, but we assume that they will remain under pressure. One reason for this assumption is that the application of biotechnology and ICT in agricultural research may result in new technologies which increase production.

Increasing agricultural productivity not seldom results in decreasing the sustainability of the farming system. For instance the Indian government has subsidised electricity for irrigation pumps. This one reason why in many areas the ground water table is falling dangerously rapidly. With increased yield more minerals are removed from the field. If these are not replenished through applying fertilizers or manure, soil fertility decreases (Sanchez and Swaminathan, 2005: 107).

For an increase in agricultural productivity many people think mainly of an increase in yields, but there are more possibilities to increase labour productivity. According to estimates of the World Bank the added value per worker in agriculture is in many Asian countries less than 1% of this value in the countries with the highest labour productivity: France and the Netherlands (World Bank, 2003: table 3.3, van den Ban, 2005). A real problem is that increased labour productivity will result in less employment opportunities in agriculture. Twenty years ago one of us was asked at the Thai Ministry of Agriculture: "Sixty percent of our labour force is employed in agriculture. If this decreases to 30% it is still possible to produce enough food for our population. Please, tell us how we can find employment for the other 30%". He could not, but fortunately some Thai people are more intelligent than he is. The employment in agriculture in Thailand has decreased to about 30% without causing large scale unemployment. However, such an increase in non-farm employment can not be realised in all other countries. As long as this large gap in labour productivity between farmers in Asia and in Europe and America will not decrease unfortunately most Asian farmers will remain poor.

Many Asian farmers search seriously for non-farm sources of income for themselves or for their children (Ellis, 2000). However, they get little help of extension agents in this search.

Globalisation of trade in agricultural products

In recent decades there has been a large increase in the international trade in agricultural products, because

- the transport costs have decreased,
- the demand for high value products, like fruits and aquaculture products, has increased,
- multi-national companies and supermarket chains play an increasing role in this trade,
- the development of ITC makes it much easier to discover where one can buy good quality products for the lowest price.
- trade barriers have decreased as a result of the WTO and they may decrease even more rapidly in the next decade as the political power of farmers in rich countries decreases with the large decrease in the number of farmers there.

A result is that Asian farmers now have to compete not only with farmers in neighbouring villages, but with farmers all over the world. To be successful in this struggle they have to be well informed about consumer demand in other countries, especially the quality they demand and they have to be served by an efficient marketing system (Swanson, 2004). Bio-safety and intellectual property rights are important in this process.

Increasing gaps in income

A result of the changes mentioned above is that in many countries the income of most farm families decreases, whereas at the same time many urban families increase their income. Many governments, e.g. the Chinese and the Indian, are worried that this increasing gap in income may cause serious social unrest. Therefore they are eager to develop policies which decrease this gap. Also many donor agencies have in the last decades neglected the support for agricultural development, but they now realise that this has to change (e.g. IFAD, 2001)

Possibilities to increase farm income

The World Bank (2005: Ch. 5) mentions 5 pathways through which farmers can increase their income:

- Intensification by using improved production technologies or investments in irrigation, fertilizers, etc.,
- Diversification by taking advantage of opportunities for high value products in the market,
- Expansion of farm size. In much of Asia this is only possible if other farmers decrease their farm size or leave agriculture,
- Complementing farm income by off farm income or by processing agricultural products,
- Exit from farming for non-farm occupations (Clark, 1957).

As a group farmers may also have some other possibilities, perhaps in partnership with people in other occupations (van Mele, 2005):

- Increasing the efficiency of input supply and marketing systems and other institutions which support agriculture,

- Increasing their power in these systems and institutions in order to decrease the gap between the price the farmer receives for his products and the price the consumer pays,
- Increasing their power in formulating and implementing government agricultural policies.

A problem is that not all farmers have the same power in their society. Often large and well educated farmers can exert most power.

Changes in Agricultural Extension

Major changes in agricultural extension we observe in Asia, partly as a result of the changes in society mentioned above, are:

- From extension provided by a government extension service to a pluralistic extension system, which includes extension by NGOs, farmer's organizations, consultants and companies providing inputs and marketing agricultural products and is financed from other sources than the taxpayers money. Often agricultural development problems cannot be solved by one of these institutions, but only by a partnership of several of them (van Mele, 2005).
- From extension which tries to increase crop yields and production per animal to extension which also helps farmers to produce products for which there is a growing demand in the market and helps farmers to increase their power in the system of input supply and marketing. This implies a change from an extension system which transfers production technologies to an extension system which helps farmers to decide what is in their situation the optimal farming system to realise their goals.

- From extension which transfers technologies developed at agricultural research institutes to farmers to extension which stimulates farmers to experiment and to learn from their own experience and the experience of their colleagues.
- From a top-down approach in extension to a participatory approach,
- Towards an extension service which tries to help to alleviate poverty. The fast majority of the poor people in Asia are small farmers and farm labourers (e.g. Christoplos and Farrington, 2004).

Sources of information for extension agents

These changes require a change in the information sources extension agents use. In the past agricultural research at government research institutes was the main source of information for extension agents, but now also many other information sources have become important, such as:

- *researchers from agri-business firms.* In several countries the agricultural research budget of these firms is now larger than that of the government,
- *market information.* For which kind and quality of products is there a growing demand in the market? In which market can one get the best price for the products or can one obtain inputs of good quality for the lowest price? What are the advantages and disadvantages of different kinds of contracts with market partners? Usually government research institutes are not the best source of information for this market information; farmers' organizations may be a better source,
- *successful farmers.* It is important to learn from farmers which criteria they use to select successful

production technologies and farming systems. Farmers may have good reasons not to follow the recommendations from their extension agents and researchers. In Europe and North America learning from the most successful farmers and teaching that knowledge to other farmers has always been a prime role of agricultural extension agents, because the experience these farmers gain in their experiments is often very relevant for neighbouring farmers.

- *government policies*. The choice of a farming system or a production technology should not only take into account the present government policies, but also the likely changes in the policies in the years to come. This may influence, e.g. the extent to which farmers will be exposed to competition with farmers in other countries.
- *other extension agents*. Successful extension agents learn from their own experience and the experience of their colleagues, just like farmers (Hall et al., 2005).

For support to decisions on the choice of farming systems other information sources are important than for the choice of agricultural production technologies. For these technologies agricultural researchers are an important source of information, although information is also needed on the resources and goals of farm families and on the reasons why they adopted or do not adopt new technologies. For farming systems, however, much of the information has to come from the markets and from the experience of farmers who experimented with these new farming systems. For deciding whether or not to change from cereal production to vegetable production, it is important to know the labour requirements on different times of the year. This information can not be collected on a research institute, but only on farms who made this change. Also market information is quite important for decisions on the choice of farming systems. This includes information on the risks involved.

A successful change in a farming system usually does not depend on decisions by an individual farmer, but requires also changes in his socio-economic environment (Leeuwis with Van den Ban, 2004). It will require changes in the input supply and marketing system and in all kinds of government and private organisations supporting agricultural development. This implies that the new farming system cannot be introduced by an individual farmers, but only by a group of farmers who like to make a similar change. For example, a trader may not be interested to come to the village to collect the product from one farmer, but only to collect the products of a group of farmers whose products can be pooled to sell them to a supermarket chain.

Advice on a change in farming system requires a different relationship between the extension agent and the farmer than on a change in production technology. On this last change it may be possible to give a recommendation, e.g. to use a new variety which is resistant to an important disease or to use a soil test to decide which kind and quantity of fertilisers to apply. A change in farming system always involves risk, e.g. on the future development of prices of inputs and products. Whether or not it is profitable for a farmer to change his/her farming system depends on his/her ability to compete with other farmers and hence on the available resources and the competence in implementing this system. The role of extension agents in developing new farming systems is not only transfer of technologies, but mainly developing entrepreneurship and facilitating the process of decision making including facilitating the development of farmers' organizations which support this new system (Heemskerk and Wenning, 2004).

Several chapters of this book discuss how farmers, researchers extension agents and others cooperate in developing technologies which work well in the specific socio-economic and agro-ecological situation of a certain group of farmers' participatory technology development (also van Mele, 2005). In this cooperation farmers may ensure that technologies are developed, which they can finance from the limited amount of capital they have.

Financing agricultural extension

In the past agricultural extension was mainly provided by government services paid by taxpayers. Now it is often said that it should be paid by the farmers themselves, because they are the people who profit from extension. In our opinion persons who say this overlook some points:

- Farmers adopt new production technologies often only after they have seen that these technologies work well on other farms in a similar situation than their own. Therefore farmers, who have learned from extension about these technologies are not the only ones who profit from this knowledge. Their friends, relatives and neighbours may profit as well,
- Technologies which result in increased production, result in decreased prices for these products. Therefore the consumers may profit more from extension than farmers, but farmers are also consumers. It may be profitable for the consumers in their role of taxpayers to pay for the extension because it contributes to a decrease in price of their food.

Crucial in agricultural innovation is the flow of information between different actors involved. This flow is influenced by the way agricultural extension is financed. For example, farmers learn from each other about the value of innovations. If a farmer had to pay the extension service for the knowledge he learned from them, he may less inclined to share this knowledge with other farmers than when this knowledge was available free. An extension agent can only be successful if he learns from farmers about their experience. But are farmers willing to share this experience with their extension agent, if they have to pay for the knowledge they learn from the extension agent, but are not paid for what they teach these extension agents?

We do not deny that there are situations in which privatisation of government agricultural extension services is in the interest of the country, but one has to analyse carefully when this is the case (van den Ban, 2000; Katz, 2002). A privatised extension service can only be profitable if it provides information and advice which farmers consider important. Otherwise they do not pay for this service, whereas some governmental extension services do not do much useful work in the opinion of many farmers. This is partly because they lack the operational funds, which they need for implementing the extension programme. It is possible that the government pays an NGO or farmers' organization to deliver an extension programme.

Use of Information and Communication Technology (ICT) in Agricultural Development

In recent years we have seen a very rapid development of information and communication technologies. "Computing power per dollar invested has risen by a factor 10.000 over the past 20 years" and "The cost of voice transmission circuits has fallen by a factor of 10.000 over the same 20 years" (World Development Report 1998/99: 57). This has resulted in a very rapid rise in the use of Internet, mobile telephones and digital camera's and related technologies. This offers large opportunities to increase the effectiveness of agricultural extension and at the same time to decrease its costs. We should discover what are the best ways to realise these opportunities. In this regard Asian countries can learn a lot from each other's experience rather than trying to invent the wheel again in each country.

It is important to learn how Information and Communication Technologies can be combined with other communication methods in an effective extension programme. In this field there is much experience in India, a world power in ICT. One such innovative example is stated below:

Ministry of Agriculture, Government of India has initiated a programme called, *Kisan (farmers) Call Centres* in which a farmer from the remote place can call over telephone to the subject matter specialist (SMS) in the district headquarters (at agricultural departments, agricultural colleges, agricultural technology information centres, etc.) and pose problems about his farming to get the suitable solutions at various levels. Besides, agricultural information KIOSKs and other service centres opened by agri- business companies and NGOs also provide the needed information to the farmers through electronic media.

Conclusion

We hope that this book will help our readers to adjust their agricultural extension system and approach to the changing situation in which they work by learning from the experience in different Asian countries. We do not provide a recipe to improve the way they work at present, but try to give ideas with which they may like to experiment together with other actors in agricultural development process which support the same group of farmers. By learning from these experiments they may develop better ideas to improve extension in their situation than we can provide. We offer a "menu for the reader to select their own dish. Ingredients can be replaced, spices added" (van Mele, 2005: 258). Sharing experiences between nations may help to improve agricultural extension and in this way to decrease poverty among farm families, because it helps them to realise the opportunities the change in their society offers them.

Editorial Overview

This book is divided into 4 editorial parts. Changes in the role of agricultural extension are clearly related to changes in extension policy. Therefore in the **First Part** those chapters are included which focus mainly on these policies. Often they do not only present the policies, but also a discussion on the way these policies have changed in the past and why they have changed.

In the past most Asian farmers were subsistence farmers who consumed in their own family most of the products they produced. Now many, not all, have become commercial farmers who sell a large proportion of their production in the market and use the money they earn in this way to finance the needs of their family, it may be school fees or even foods they do not produce themselves. They do not only need the information on production technologies, but also market information as is discussed in three chapters in **Part 2** on

Commercialization. This implies that a change in the role of extension is essential as is discussed in three chapters of this Part.

Not only farmers become commercialized, but also extension services. They are no longer completely provided by the government, but also through the money they earn by giving information to farmers or providing education to them. Therefore this part of the book also includes a chapter on the privatization of extension.

Most poor people in Asia live in the families of small farmers and farm labourers. In nearly all countries a major goal of the agricultural extension services of the government and the NGOs is to reduce their poverty. For many of these families commercialization may not be the best solution, because that it too risky and they may have difficulties to compete in the market. The experiences with this kind of extension are discussed in three chapters in **Part 3.**

Nowadays farmers have not only to compete with other farmers in their own village, but with farmers all over the world. Their managerial competence determines to a large extent how successful they will be in this competition. Many see it no longer as the main role of extension service to transfer information about production technologies from farmers, but more to increase through education the ability of farmers to make their own management decisions as is discussed in **Part 4**. It can be important to help farmers to make a decision about a change in their farming system, because changes in the economic environment and technological development may make a different farming system more profitable. In this Part also two chapters are included which discuss the experiences with the use of

Information and Communication Technologies (ICT) for educating farmers and transferring information to them.

Part 1 : Extension policies

Five Chapters are included in this Part. It starts with a chapter by Sulaiman and Hall, which analyses these policies in different Asian countries and gives attention to the process of developing an extension policy. An extension policy is only effective if it serves the need for support of the target group. Therefore, this group should be able to influence policy decisions. With the rapid change in Asian societies agricultural extension has to change as well, but these changes can not be realized by decisions at the top. The people who have to implement these decisions should also be involved in the decision making process.

The case from Korea is interesting because this country started already 40 years ago with a rapid economic growth. Can we expect that several of the developments in Korean agriculture, which are discussed in Ch. 3, will in the next decades also happen in other Asian countries, e.g. the rapid decrease in the farm population and the development of specialized farms? Dwarakinath has for many decades been involved in designing and implementing extension policies of the government and NGOs in India. His experience is discussed in Ch. 4. Israeli farmers have now a high level of education and much power to influence agricultural development and extension policies, as is discussed in the Ch. 5 by Blum.

The discussion of T & V Extension in Ch. 6 gives an interesting insight in the decision making an donor financed extension projects based on the experience of the World Bank.

Part 2 : Commercialization of farming and privatization of extension

As a result of the rapid commercialization of farming in many Asian countries, farmers now need a very different support from extension than in the time that most of them were subsistence farmers. They need more help in marketing and in deciding how to change their farming system. This process is discussed by Hasannullah in Ch. 7, mainly based on his experience in Bangladesh. Singh, Swanson and Singh describe in Ch. 8 a World Bank Extension Project they managed, which tries to help Indian farmers to increase their income by profiting from the opportunities in the national and international market to sell high value products. Information on how to do this comes from farmers, who have done this already successfully, researchers with many different specializations, input supply and marketing companies, extension agents in different government agencies and NGOs. Therefore one has to built an organization in which many different stakeholders cooperate.

In Ch. 9 Linh describes how one works in Vietnam on the assumption that extension will be successful if it provides information for which farmers feel a need. Therefore, in many projects one has studied these information needs. A conclusion is that farmers are often more in need for information on the marketing of their products than on production technologies. This extension approach requires that the extension agents themselves are well informed about marketing.

Qamar discusses in Ch. 10 in which situations a privatized agricultural extension service can be successful. But also the dangers of privatizing a government extension service in situations where no financially viable private extension organization can be developed. This is a topic which is often not mentioned by donors who give recommendations about privatization, but in working for the FAO he has seen in many countries that these dangers are quite real.

Part 3 : Supporting poor farm families

The people who need support from agricultural extension most are the poor farmers, who often live in risk prone, remote areas. Often they receive least support, because many extension agents and agricultural researchers do not know how they can improve their situation and the farmers do not have the political power to press for support. Rangnekar has devoted most of his live in supporting them in a large NGO and with the Indian Dairy Development Board. Ch. 11 is based on this experience.

Opatpatanakit is doing this in Thailand by involving these farmers themselves in research to develop solutions for their problems. They are more inclined to apply solutions they have discovered themselves than solutions which researchers or policy makers tell them that they should use (Ch. 12). Biradar works with dry land farmers in India along a similar line, but develops a process in which knowledge and experience from farmers, researchers and extension agents is integrated (Ch. 13). As a staff member of a research institute it is her experience that this kind of action research helps more to give researchers a good understanding of the problems, situation and possibilities of poor farmers than their participation in a PRA. This understanding is required to enable them to do research which contributes as much as possible to poverty alleviation.

Part 4 : Increasing managerial abilities of farmers and the use of ICT

Managerial abilities are crucial to enable farmers to compete in the World Market. Based on some 15 years experience with Farmer Field Schools for Integrated Pest Management van de Fliert discusses in Ch. 14 how these schools can enhance the capabilities of farmers to make decision on cultivation practices based on the actual situation in the field and not on prescriptions they receive from researchers.

Already for a century the budget for vocational agricultural education of European governments is larger than that for agricultural extension. Mancebo shows in Ch. 15 based on the experience in the Philippines that this kind of education can also be introduced in Asia to make an important contribution to agricultural development. But it has to be combined with lifelong learning in which the extension service teaches farmers how they can adjust the management of their farm to changes in the environment and to new opportunities.

There is Chapter on the use of ICT in China by Gao and Li and in India by Ramachander and Jhunjhunwala, The Chinese chapter discusses how ICT can increase efficiency of an agricultural information service with 180 000 staff members and how valuable different information services are for farmers. It is still difficult to provide through ICT the location and situation specific information farmers need. The Indian chapter gives now much attention to the way the government extension service uses ICT, but more how commercial companies use it in their transactions with farmers on inputs and agricultural products and how kiosk owners earn a living by providing farmers access to information on ICT networks. It discusses also the rapidly changing technologies through which the villagers can be reached.

Both chapters do not discuss what can be the optimal role of ICT and of other information sources in the decision making process of the farmers. That is an issue which requires further research.

References

Christoplos, I., and J. Farrington, eds. (2004). *Poverty, vulnerability and agricultural extension: Policy reforms in a globalizing world.* New Delhi, Oxford University Press.

Clark, C. (1957). *The conditions of economic progress.* London, MacMillan, 3rd ed.

Ellis, F. (2000). *Rural livelihoods and diversity in developing countries.* Oxford, Oxford University Press.

FAO (2002). *World Agriculture: towards 2015/2030.* Rome, FAO.

Hall, A. et al. (2005). Institutional learning and change: a review of concepts and principles. *Policy Brief 21, National Centre for Agricultural Economics and Policy Research*, New Delhi.

Heemskerk, W., and B. Wennink (2004). *Building social capital for agricultural innovation: Experiences with farmer groups in Sub-Saharan Africa.* Amsterdam, Royal Tropical Institute.

IFAD (2001). *Rural Poverty Report 2001: The challenge of ending rural poverty.* Oxford, Oxford University Press.

Katz, E. (2002). *Innovative approaches to financing extension for agriculture and natural resource management.* Lindau, LBL, Swiss Center for Agricultural Extension.

Leeuwis, C., with A.W. van den Ban (2004). *Communication for rural innovation: Rethinking agricultural extension,* 3rd ed., Oxford, Blackwell Publishing.

Sanchez P., M.S. Swaminathan et al. (2005). *Halving hunger: it can be done. UN Millennium Report Task Force on Hunger.* London, Earthscan.

Swanson, B.E. (2004). Extension strategies for poverty alleviation in a global economy, in: *2nd International Conference on Agriculture Education and Environment, Best practices in poverty alleviation through education in the new century.* Asia Pacific Association of Educators in Agriculture and Environment.

Van den Ban, A.W. (2000). Different ways of financing agricultural extension. *AgREN Network Paper 106b*. London, ODI.

Van den Ban, A.W. (2005). Changes in society require changes in agricultural extension: Some observations. *Asian Journal of Extension Education, XXIII: 1-16.*

Van Mele, P., A. Salahuddin and N.P. Magor, eds. (2005). *Innovations in rural extension: Case studies from Bangladesh*. Wallingford, CABI Publishing.

World Bank (1999). *World Development Report 1998/99; Knowledge for Development.* New York, Oxford University Press.

World Bank, (2003). *World Development Indicators*. Washington, D.C., World Bank.

World Bank (2005). *Agricultural growth for the poor: An agenda for development*. Washington D.C., World Bank.

Part 1

Extension Policies

2

Extension Policy Analysis in Asian Nations

V.Rasheed Sulaiman and Andy J.Hall

Introduction

Agricultural extension in Asia, particularly in low-income countries, is struggling to reinvent itself. For decades the policy emphasis has been on public sector provision of services to extend new technologies to farmers. Public extension has and will continue to play an important role in most Asian countries. Without public funds for extension, substantial public interests are compromised especially those concerned with ecological sustainability and poverty reduction (Katz, 2002). However there is growing recognition that a narrowly defined model of public provision of technology transfer services has outlived its usefulness as an effective agricultural development strategy. The agricultural sector undoubtedly remains important in most Asian countries, but the nature of agriculture and the rural sector more generally is changing and this is placing new and more complex demands on extension.

These new challenges mean that extension and extension policy need to tackle diverse objectives that include but go beyond transferring new technology. These include: the need to link more effectively and responsively to domestic and international markets where globalisation is increasing competition; the need to reduce the vulnerability and enhance voice of the rural poor (Farrington *et al*, 2002) and promote environmental conservation (Alex et al 2002); the need to view agriculture as part of a wider set of rural development processes that include enterprise development and non-farm employment (Rivera 2001); the need to couple technology transfer with other services relating to both input and output markets (Neuchatel Group, 2002); and the need for a capacity development role for extension that includes training but also includes strengthening innovation process, building linkages between farmers and other agencies, as well as institutional and organisational development to support the bargaining position of farmers (Sulaiman and Hall 2003). The last decade has also seen the almost universal questioning of the most appropriate role of the state in the provision and financing of services. At the same time alternative and complimentary service providers from the private and non-government sectors are emerging. It is within these broad contours that extension policy in Asia needs to be considered.

Extension and innovation: an emergent paradigm?

While extension was originally conceived as a way of transferring technology to farmers there is now wide recognition that this task needs to expand considerably (Sulaiman and Hall 2003, Rivera, 2003, Alex et al, 2002). In the same way the earlier reliance on the development of extension models that could be widely replicated across countries and regions has proved to be ineffective. There is an increasing realisation that new extension approaches need to emerge locally, based on experimentation, learning and adaptation to prevailing circumstances (Echeverria, 2003, Sulaiman and Hall, 2003).This point of view emerges from a growing understanding of the nature of technical change and the recognition of the complexity underpinning innovation processes. Modern theories of innovation suggest that technology and knowledge more generally is often intimately related to its context of development and application.(Gibbons *et al*, 1994) While generic crop production technologies can be developed centrally, it is often their local adaptation that allows technological change and innovation to take place. Modern innovation theories also point to the fact that technology development and use are often embedded in a wider set of complimentary activities and relationships. These are often related to the market and involve complimentary inputs such as credit or other forms of knowledge. What this suggests is that ways of promoting agricultural innovation require processes and relationships to adapt and complement new technologies in particular locations. In the same way this locally relevant learning and knowledge creation may take many different forms depending on local circumstances and these may change over time as new challenges and opportunities arise. These ideas are increasingly being discussed in terms of an innovation system.

This view has many attractions given the complex and dynamic challenges of contemporary rural development scenarios. A significant implication is the stronger capacity development role that this perspective suggests for extension. And in this regard capacity development does not relate to training farmers, but that to knitting together the relationships and partners needed to support local learning and innovation. Such an approach suggests incremental and evolving development of extension approaches and the possibility of great diversity emerging. This doesn't only suggest different extension policies are required. It also suggests a different policy process where by policy provides a facilitating framework and principles for local experimentation and learning. Such experiences then inform policy about adjustments that need to be made to principles and facilitating frameworks. Of course this contrasts sharply with the conventional extension policy processes where policy is a centrally generated plan with detailed prescriptions for implementation at the local level. While a more interactive approach to extension policy seems to have many merits, Sulaiman and Hall (2003) have cautioned that it is a significant challenge for extension services and associated bureaucracies to adjust their professional cultures to embrace diversity, autonomy and learning and this is the central challenge for extension policy today. How then are extension policies in Asia faring in these challenging times?

Agriculture and extension arrangements in the Asian Region

The rural population of Asia, one third of the worlds total, depends almost entirely on three activities: agriculture, forestry and fisheries. Dependence on agriculture is significantly higher for more than half of the countries in the region. Most of the Asian countries initiated measures to strengthen domestic food production during the last 50 years, many after attaining independence. The Central Ministry of Agriculture in each of these countries had taken the lead in planning, financing and implementing strategies for agricultural development. This has often been in

Table 1 : Major patterns of agricultural and rural development in the Asian Sub-region

Zones	Countries	Importance of agriculture	Patterns of agriculture and rural development
South Asia	Afghanistan Bangladesh Bhutan India Maldives Nepal Pakistan Srilanka	Agriculture contributes a significant share of GDP in this region and provide employment to more than 50% of the work force throughout the region. Dependence in agriculture significantly high in India, Bangladesh, Pakistan and Nepal and Srilanka.	Greater number of undernourished and poor than any other developing region. Highest rural population density, predominantly small farm agriculture, significant increase in production and productivity of food crops in irrigated regions. Except Afghanistan, all other countries have the necessary extension infrastructure. T and V extension was implemented in Bangladesh, India, Nepal, Pakistan and Srilanka
South East Asia	Brunei, Cambodia, Indonesia Laos Malaysia Myanmar Philippines Singapore Thailand Vietnam	Agriculture employs large proportion of the work force in all countries except Singapore. Dependence on agriculture is significantly high in Cambodia, Laos, Myanmar and Vietnam.	Indonesia and Malaysia have a well established plantation sector comprising large estates. Rice is the most important crop and has seen significant production and productivity increase. Extension infrastructure relatively well developed in Malaysia, Indonesia, Thailand and Cambodia. T and V extension was implemented in Indonesia, Malaysia, Thailand and Philippines.

Zones	Countries	Importance of agriculture	Patterns of agriculture and rural development
East Asia	China Hong Kong Japan Korea DPR Korea Rep Mongolia Taiwan Timor	Except Hong Kong, agriculture is an important economic activity for countries in the region. Agriculture employs large proportion of the work force in China, Korea DPR, Mongolia and Timor.	Japan and Korea Rep are the most agriculturally advanced countries in the region. A well-established extension infrastructure exists in Japan, Korea Rep, Taiwan and China.
Central Asia	Kazakhstan Kyrgystan Russian Federation Tajikistan Turkmenistan Uzbekistan	Agriculture employs a large proportion of the population in this region.	Countries in a complex process of transforming their political and economic systems. Lack an organised and technically competent extension system and steps are currently being taken to establish them -
West Asia	United Arab Emirates Bahrain Iran, Iraq, Israel Jordan Kuwait Lebanon, Oman Palestine, Qatar Saudi-Arabia Syria, Turkey Yemen	Agriculture employs a significant proportion of the population in Iran, Palestine, Syria, Turkey and Yemen Though employs around 2% of its workforce in agriculture, agriculture is an important item of commerce for Israel.	Israel has the most technically advanced agriculture in the region and has one of the very successful extension system. Yemen implemented a T and V approach of extension for five years starting 1985.

served by the public sector and in addition are weakly integrated into the market with limited private sector activity.

Thirdly, pluralistic institutional arrangements are emerging and this is finding wider acceptance everywhere. Farmer associations are equal partners in extension in countries such as South Korea and Taiwan. In Israel, farmers even "contract-in" certain services. China is currently encouraging constitution of farmer associations to take up various production, marketing and extension functions. NGOs and the private sector play an important extension role in India, Bangladesh, Malaysia and Sri Lanka. Extension provision by private companies to farmers growing crops under contract is gaining importance in China and India.

Fourthly, countries increasingly realise the need for extension to engage with a wide range of issues beyond disseminating technologies. This has raised the need for better-qualified and specialised extension staff to meet the changing information and technical demands of farmers. Similarly it is now recognized that there is need for extension to play a greater adaptive research role to better target technologies at the field level and to provide organisational and marketing support to farmers. However playing this wider role requires large-scale restructuring and institutional change, which, by and large, the extension bureaucracies have been reluctant to undertake. Reinforcing this reluctance is an extension policy dialogue that continues to be couched in terms of a narrow conceptualisation of extension as an agency transferring technology and improved practices from research stations to farmers.

Fifthly, not all countries in the region have an explicit extension policy. India has a Policy Framework for Agricultural Extension (DAC, 2000) and Bangladesh has a New Agricultural Extension Policy (DAE, 1999) . But the available evidence indicates that having an extension policy is only a necessary but not a sufficient condition to guide change. Policies also need to address the crucial problems involved in implementing change (Sulaiman and Hall, 2002). Quite often, policy related to extension stems from changes in country development plans (Malaysia, Indonesia, Iran), donor interests (T&V), changes in agricultural and rural policy (China, Vietnam) or change in governments (Pakistan, Philippines). Extension services have always tended to respond to changes taken by other systems and have rarely guided changes in extension or rural development policy.

Agricultural Extension Policy and Policy Process in the region

The FAO Global Consultation on Agricultural Extension (1990) recommended that "all national governments should develop and periodically review their agricultural extension policy. The policy should include the goals of agricultural extension, the responsible agencies and personnel, the clientele to be served, the broad programmatic area to be addressed and other relevant guidelines. In developing national agricultural extension policies, representatives of all major groups of farmers should be directly involved and other relevant agricultural organisations should be consulted. By pursuing a comprehensive policy, countries can expect the extension system to contribute to increasing agricultural productivity and farm incomes, and to improving the quality of life of most rural farm households in pursuit of the general goal of growth with equity" (Swanson, 1990). However most of the countries have not taken this recommendation seriously and even after a decade, only a few governments such as India and Bangladesh have attempted to develop a formal extension policy.

Extension policy changes are often made through decrees and proclamations and only in rare cases, it is legislated. For instance, Japan enacted the Japanese Agricultural Promotion Law of 1948 and provided funding for Japan's Co-operative Agricultural Extension Service. Similarly Agricultural Extension Policy in South Korea is embodied in the 1957 Agricultural Extension Law and in the Rural Development Law of 1962 (Contado, 1997). After the end of external funding for the T&V system, several countries made ad-hoc changes in extension approaches mainly to tide over the increasing financial liability of increased manpower the T &V created and also to reach directly to more farmers. In India, states responded in various ways depending on the local situation (Sulaiman, 2003) and this process continues even though there is a policy framework for agricultural extension at the national level. Bangladesh adopted the New Agricultural Extension Policy in 1996. Of particular emphasis in both the Indian and Bangladesh extension policies is the emphasis on promoting partnership among all extension providers, namely the GOs, NGOs and the private sector. However, lack of a shared understanding among the central and lower management levels on the importance of new agricultural extension policy and the inability to make the necessary cultural change currently constrain the development of a partnership mode in extension (Uddin, 2003).

In the case of Malaysia, the Third National Agricultural Policy (1998-2010) has given the impetus for change in approaches and functions. Keeping in view the goals of maximisation of income, a greater role for the private sector and increased competitiveness of

Malaysian agriculture, the mission statement of the Department of Agriculture currently includes, provision of quality advisory and consultancy services to farmers, entrepreneurs and private sector. A private-public co-ordination council has been established at the government level to plan and co-ordinate activities.

In Vietnam economic reforms known as "doi moi" have shifted power away from central authorities towards groups of more autonomous actors. Extension in Vietnam is relatively a new phenomenon, embarked upon in 1993. It was associated with the land reform process which changed land tenure from collective to household ownership. The formation of policy and an organisation for extension is therefore gradually developing and finding its form. Ministry policy documents mention the objective to gradually move towards cost-sharing of extension services for commercial production, and subsidised services in the remote mountain areas (Farrington, et al, 2002).

In Central Asia, the countries, which exercised socialist policies for many years are struggling to shift to a market oriented economy. Thousands of new farmers who currently own private land are desperately in need of extension services. To exploit their full potential in agriculture, appropriate national extension systems have to be established through institutional reforms and backed by national policies (Qamar, 2002).

Since the beginning of the 1980s, a rural reform policy has been initiated in China. This has been part of fundamental economic reforms that China has undertaken, particularly the opening up of the economy to global markets in the last 20 years or so. The mission and goals of rural extension were reinvented in China to meet the changed situation and China is perhaps the only country in the region that is expanding its extension infrastructure and experimenting with several new strategies. Indonesia shifted its paradigms of agricultural development in response to the second 25 year Development Plan that began in 1993. To revitalise small-scale farming and to realise economies of scale, a group approach to farming is followed at present. Based on Public Law No.22/1999, the district governments are directly responsible for planning, financing and implementation of extension activities at the district level. In Pakistan, based on the devolution plan of 2001, all service delivery line departments including agricultural extension were transferred to the district governments. Devolution has marginalized extension services in both Indonesia and Pakistan. In Sri Lanka, responsibility of agricultural extension in non-plantation agriculture was devolved to provincial councils in 1989. The Extension, Education and Communication Service (SDRE) of the Food and Agricultural Organisation (FAO) currently assists Yemen, Indonesia, Iran and Philippines to help them establish and/or strengthen their devolved extension services (Rivera, *et al*, 2001).

In the case of Iran, the status and role of agricultural extension is still a matter of dispute due to lack of a clear policy on what it has to do and which could be the ideal ministry or organisation in the government that extension should be linked to. In Israel, the national extension service, SHAHAM, continues to be operated as an arm of the Ministry of Agriculture and Rural Development, although it has introduced increasingly more elements of privatisation and financing by users. Agricultural extension service fulfils advisory and applied research functions and provides technical consulting services to the management of the Ministry of Agriculture in formulation of its agricultural policy.

Except the Israeli extension service, no other extension service play any role in formulating the country's agricultural policy. Quite often extension was found to simply respond to changes imposed on its functioning. The private sector, farmer's organisations and the NGOs are not given any role in the formulation of extension policies and programmes in most Asian countries. However mechanisms for regular consultations with NGOs and farmer representatives on issues related to agriculture and fisheries have been created in the Philippines. The (Philippine) National Agricultural and Fisheries Council (NAFC) is an inter-sectoral and inter-agency body having representation from peoples organisation and NGOs engaged in agriculture and fisheries sector. NAFC facilitates regular consultations and dialogues between government and the private sector. Agricultural and Fishery Councils have been set up at the sectoral, regional, provincial and municipal levels, involving farmers, fishers, traders, rural bankers and agri-business entrepreneurs to provide inputs on major programmes and policy decisions and help plan and monitor programmes.

In order to illustrate some of these development and challenges in extension policy the next section provides a more detailed discussion of the situation in 4 major countries in this region, namely, **India, China, Indonesia** and **Iran.**

India

Agriculture contributes about 24% of the GDP and employs about 57% of the total workforce in India. Every state in India has the authority to legislate on matters related to Agriculture. In terms of number of staff and organisational reach, the public sector state Department of Agriculture (DoA) continues to dominate extension provision. A large number of private agencies provide advisory and other support services to farmers. Most of them are restricted to certain regions and selected crops. However public-private partnership is very limited. There are very few farmer organisations. Technology dissemination continues to be understood as the main extension role. The post T&V period saw several states introducing: decentralisation (extension planning and control under elected bodies at the district/ block level; contracting NGOs for some extension activities, adoption of a group approach (instead of the earlier individual approach); use of para-extension workers (as substitutes for field extension workers of the DoA; setting up multi-disciplinary teams from the State Agricultural Universities at the district level; setting up of agri-clinics by private entrepreneurs; and formation of registered society known as the Agricultural Technology Management Agency (ATMA) at the district level by integrating the functions of key local stakeholders involved in agricultural development. However the DoA still face several constraints in providing adequate extension support to farmers. Remote areas and poor producers (especially those growing low value crops and having limited marketable surplus) are poorly served by both private as well as public sector extension.

The Ministry of Agriculture, Government of India brought out a draft Policy Framework for Agricultural Extension (PFAE) in 2000. The Ministry initiated consultations with the state governments and the private sector on this draft document. Highlighted are the following measures suggested in the PFAE : adoption of farming systems and farmer participatory approaches; enabling problem solving skills of farmers; provision of public funds for private extension; privatisation of *private-good* elements of extension in favourable areas; provision of cost recovery and co-financing via farmer groups; use of para extension workers and farmer interest groups in extension; employing more subject matter specialists;, single window services at block (an aggregate of about 100 villages) level; preparation of strategic research and extension plans; improvement of research-extension-farmer interface; improvement in women's access to technology; provision of market information; wider use of information technology; promotion of linkages with agro-processors and creation of an enabling environment for private sector.

Though the broad contours of policy changes suggested are well considered and relevant, the PFAE underplays crucial implementation problems of introducing reforms (Sulaiman and Hall, 2002). Of particular concern is the wider institutional framework of public extension and the restrictions this imposes on the introduction of change and the development of new approaches. A central concern is the current organisational culture of the extension service and associated issues related with public sector governance in general in India. This include; rigid professional hierarchies and patterns of control, with highly centralised modes of planning which stifle deviation from prescribed procedures and restricting innovation (both technological and institutional) by middle and lower level staff; a tradition of assessing performance in technology adoption and hence a focus on improved technology transfer at the expense of other activities that may have a perfectly legitimate role in supporting farmer; a history of rewarding successes and thus a reluctance to report and analyse the reasons for failure for, instance, non-adoption of technology; a history of working independently and a mistrust of other agencies; and a tradition of up-ward accountability for resource utilisation rather than output achievement and client satisfaction. There has been no attempt to address these crucial issues. The reforms suggested in the PFAE will not by themselves reinvent Indian agricultural extension unless they are accompanied by a much more explicit agenda of institutional learning and change for the public agencies involved (Sulaiman and Hall, 2003). The case of India suggests that there is a need to close the gap between policy prescription and policy implementation. The policy process probably needs to take stock of recent developments and experiments in the area of extension and think about ways in which it can create stronger functional links between ground level realities and policy formulation.

China

Agriculture accounts for about 15% of the GDP and employs 66% of the total work force in China. Since the beginning of the 1980s a rural reform policy has been initiated. First, collectively owned land resources were contracted to individual households for long periods (more than 30 years) and farmers now have right to decide how to use the land. Government has also gradually reduced its control of the agricultural market. These reforms brought new challenges to the extension system. Since the market oriented economic system was established, the subjects of rural extension have been expanded and diversified according to local resource and market development. Current functions include: transfer of technologies; providing market information; creating market organisations; and assisting purchase of production inputs and promoting rural enterprises. Extension operates through technical demonstration households, village leaders, farmer associations and farm households. Since 1985, the central and provincial governments have invested about 1.2 billion Yuan (145 million US $) for the construction of county centres and the number of extension staff paid by the government and working at the township level has reached 1 million (Yonggong,1998).

Apart from the conventional extension approach that implements government plan through public extension agents, China employs other innovative extension approaches, such as technical contracts between farmers and extension agents, the (private) company led extension approach and participatory extension approaches. Technical contract is signed between extension agents and farmer households and under this approach the extension agent provides technical training, on-site supervision and instruction during the production period. The approach is mostly applied in horticulture, cash crop production and livestock production systems. The agent and the farm household share profits and risks under this arrangement. Under the company led extension approach, companies provide relevant technologies, training and information to farmers who enter into contract farming arrangements. To develop farmers' abilities and skills in sustainable rural development, participatory extension programme has been introduced. Community is the basic unit for implementing participatory extension programmes and farmers participate in all extension processes such as project appraisal, monitoring and evaluation. Extension priorities have been identified for the 7 different geographical regions (based on agro-resource planning) and for the three economic zones (based on average income and GNP).

Extension policy at the national level is formulated by the national extension institution, the National

Agricultural Extension Centre working under the Ministry of Agriculture. This centre draws up extension programmes that link agricultural development programmes, connecting institutions with other national agencies and training and supervising provincial agents. Provincial institutions under the Department of Agriculture, act as professional agencies of the national extension centre and these centres are involved in formulation of policy, co-ordination of relevant agencies and training of lower level agents. There is a high level of governmental policy orientation and limited participation of farmers in the planning, implementation and evaluation of extension programmes. Recent analysis (Yonggong, 1998) suggests that some of the weaknesses of the Chinese extension system include: insufficient qualifications of extension staff; inadequate professional and institutional linkages between research and extension; and insufficient co-operation and co-ordination with other agencies involved in planning and credit. While the Chinese extension policy is weak, implementation and innovation in extension approaches at the local level is strong. Experimenting with such diversity of approaches would not have been possible within a centrally decided extension policy for all regions. This indicates the need for facilitating the local development of capacity to innovate appropriate policies and approaches.

Indonesia

Agriculture contributes approximately 35% of the total GDP and currently employs about 43% of the working force in Indonesia. Agriculture has a dualistic structure, i.e. small-scale agriculture involving nearly 24 million farm-households covering about 142 million ha and plantation agriculture operated by about 2000 estates covering approximately 22 million ha. Under the second 25 year Development Plan which began in 1993, Indonesia moved from centralised planning to decentralised planning. Emphasis shifted from increasing agricultural production to increasing farmers income; from production of primary commodities to agri-business in rural areas; from labour intensive technology to capital intensive technologies; from import substitution to one led by agricultural exports; and from a dominant government role in economic development activities to greater participation by the private sector.

The Ministry of Agriculture is focussing on creating a policy environment to foster private sector development and market efficiency. Steps to revitalise small-scale farming are being taken by introducing an agri-business approach to a commodity based farming systems. The approach is to identify a primary commodity or produce based on market preference and co-ordinate groups of neighbouring farmers and makes them into groups to realise economies of scale (Suryana and Erwidodo, 1996). To address the research-extension linkage issues, Indonesia has established new institutions called Agricultural Technology Assessment Institute at the provincial level to bring together farmers, researchers and extension specialists (Kadir *et al*, 2003).

Law No 22 on Government Autonomy describes the responsibility and authority of central government, provincial government, as well as district government in implementing and in managing extension. The authorities decentralised to district governments are related to the planning and implementation of extension activities and provision of direct services to farmers and the community. Decentralised management of the agricultural extension programme predicted the following potential benefits: an opportunity for districts to select and decide on the activities to meet the needs of local farmers; and a mechanism to increase farmers' confidence in implementing programmes by farmers themselves. However during implementation it has emerged that the agricultural extension institution was not fully prepared to manage decentralised extension (Zakaria, 2003). Many members of the district council do not have correct or complete understanding of the concepts and roles of extension and agriculture development.

The responsibility to provide fund for salary and operational activities become the responsibility of the district government. Due to limited district government fund, many districts cannot provide sufficient fund to pay salary and operation costs of the extensionists. The allocation of available funds in the district depend on the political struggle and conflicting of interests of the members of the district council. At the same time the, the political voice of farmers is still weak. The success of decentralised extension implementation depends on the of the way the participatory extension concept and methodology is perceived and understood by members of district councils, district governments, top officials, officers, private sector, farmers and non-government organisation, There is a need for intensive socialisation of decentralised extension policy (Zakaria, 2003). So in one sense decentralisation, while part of a process of introducing more democratic modes of public planning and service provision, also needs to be supported by the cultural and institutional changes at the local level required to accommodate such approaches. Extension policy clearly needs to facilitate these changes and incremental developments.

Iran

Agriculture contributes 26.6% of GDP and 26.6% of the workforce is engaged in agriculture in Iran. The Department of Agricultural Extension became a part of different organisations and Ministries during the last five decades. In 1953, the integrated institution of agricultural extension was formed within the framework of an independent instructional organisation under the auspices of the Ministry of Agriculture. In 1973, the implementation of projects to increase agricultural production was entrusted to the Agricultural Extension Department. This diluted its earlier educational and training role as it began to focus more on the provision of loans and credits, preparation and arrangement of works, and handling distribution of seeds. Following the objective of developing a meaningful relation between research and extension, the agricultural Extension Organisation was merged with the Agricultural Research and Education Organisation in 1993 and the newly formed establishment was named Agricultural Research, Education and Extension Organisation(AREEO).

In 1978, after the culmination of the Islamic revolution a chain of establishments including the organisation of Jihad-e-Sazandegi was founded. The main objective of creating this organisation was to provide preliminary assistance for farmers and the rural community across the country and to rehabilitate and develop the rural areas. The Ministry of Jihad-e-Sazandegi implements its programmes through the Rural Islamic Councils (which are supervised by the Minsitry of Interior), and work through Rural co-helpers who are elected by the Islamic council as facilitators for rural development. This organisation was upgraded to the level of a Ministry in 1990. Thus there were two

Ministries that were involved in agriculture in Iran until 2000; the Ministry of Agriculture that was mainly involved in agronomy, horticulture and related research, education and extension; and the Ministry of Jihad-e-Sazandegi that was in-charge of animal husbandry, fishery and natural resource management. In 2000, due to problems caused by the separation of two Ministries, it was decided to re-merge the two Ministries into an integrated Ministry of "Jihad-e-Agriculture" or Agricultural Jihad. This was done with the primary objective of downsizing governmental organisational structures;, integrating plans and policies related to agricultural and rural development; and to undertake more projects and programmes by provincial and regional levels for decentralisation purposes (Heidary, 2001)

One of the major problems affecting agricultural extension in Iran is the ambiguity in its position (Gharehyazi *et al*, 2003). Under AREEO, extension was linked with research and it has been doing more of professional extension and transfer of findings. Under the Ministry of Jihad-e-Sazandegi, extension activities were not combined and integrated with research and education In fact, such activities were carried out under the supervision of Deputy Minister for Extension and Peoples' Participation. After the merging of the two Ministries, extension has become a matter of dispute. The issue of whether extension should be completely integrated with research or separated from it it is yet to be fully resolved. Serious budget shortages and in particular lack of funds for hiring extension personnel and minimal access to transportation facilities are also notable.

The present confusion on where extension should be attached may not be fully resolved as long as the role and functions of extension remains undefined. Clearly it has roots in education, research and rural development, but locating it in any of these three camps will fail. Becoming a part of the wider rural development sector can potentially help extension to broaden its mandate. To play a meaningful role, extension should ideally build relations with a wide range of actors in development and research-extension linkages is only one among them. Quite often, policy changes happen all of a sudden with total change in roles and functions implemented throughout the country. Ideally policy reforms should be a continuous process with experimentation of different kinds of approaches in limited scale and learning from these changes should inform the development of policy. Lack of such an approach is quite evident in this case.

Emerging issues in extension policy in Asia

Extension is clearly facing challenging times in Asia. The country studies illustrate four different and informative modes of extension and extension policy development. In the case of India, like a number of other Asian countries, extension policy is developed centrally in a fairly prescriptive fashion. Although approaches have evolved over the long term, it is not clear how implementation experience and learning informs policy development. In fact development fads and encouragement from international agencies seem to be a major source of implementation. From T&V, the approaches have now shifted to pluralism in extension financing and service delivery. While these new objectives might be laudable at a general level, ways of making them work on the ground is much harder to define. Furthermore these major shifts tend to lock up the extension in a particular mode of operation until yet another new idea comes along. For countries suffering from this policy trend, approaches that encourage learning and incremental institutional change are long over due.

The case of China illustrates a quite different approach to policy and practice in extension In China the challenges of major economic and indeed social reforms over the last 20 years have galvanised local agencies associated with extension to respond in flexible and meaningful ways. Arrangements have been restructured to help farmers relate to new market opportunities more effectively. Arrangements have even been developed to provide incentives to extension workers through profit sharing with farmers. The extension policy on the other hand has been fairly haphazard and weak in prescribing what should be done. But it does seem to have been sufficiently reactive to provide the facilitating support to assist and presumably legitimise the types of pragmatic extension innovations that have been taking place at the local level. There seems to be much merit in an extension policy process in which extension agents, farmers and others in the rural areas drive the development of new ways of approaching the topic. This seems to be particularly important in situations where the economic and social context is changing rapidly and in unpredictable ways.

The case of Indonesia again illustrates a trend that many Asian countries are following, namely decentralisation. The case highlights the gap between the broad policy prescription for an approach such as decentralisation and the reality of how these approaches can break down in implementation. While the policy was originally (and laudably) conceived as a way of devolving authority and decision making to local stakeholders and strengthening linkages in local knowledge networks, neither this vision nor the skills to implement it were shared by those at the local level. This seems to suggest that policy instruments such as decentralisation need to be accompanied by capacity development. In this case local stakeholders need to understand the importance and rationale for strengthening local network. And since the performance of extension is dependent on these systems, stakeholders need to have the skills to analyse these systems, diagnose system failure and design remedial measures. These types of capacity development are not only necessary to successfully implement these decentralised approaches, but they are also necessary if local stakeholders are to play a more interactive role the policy process.

The case of Iran is perhaps not as unusual as it seems. Many countries in Asia have undergone radical social change with the attendant process of fundamental changes in the arrangement of government departments such as extension. Perhaps what is even more familiar is the way extension has been passed between agencies responsible for education, research and rural development. The trouble is that extension belongs in all three and there is practically no way of reconciling a policy debate about which disciplinary or sectoral structure it bests fits into. The possible way to resolve this seems to be for policy to start and think about extension in a more systems orientated fashion, working towards breaking down some of the disciplinary and functional distinction between extension and related responsibilities of government. As seems to be the case in so many instances, ways of doing this can't really be prescribed, but will need to be approached experimentally.

Given the above it seems that extension policy needs to tackle two major sets of issues. The first concerns the content of extension policy and the urgent need to redefine the role and form that modern extension arrangements should play in the contemporary development scenario. The second issue concerns the nature of the policy process associated with extension and the increasingly untenable approach whereby policy prescriptions (sometimes responding to international fads) are disconnected from ground realities and implementation challenges.

Policy challenges

The need for extension reforms in Asian countries is widely acknowledged. Driving this is the need to meet the diversity of objectives relevant to contemporary agriculture and rural development. Without reform, public extension services in Asia will become ever more irrelevant and will lose the political support needed to fund them. The new vision of extension has to pay much more attention to ways of addressing the welfare needs of farmers in rapidly changing rural scenarios. This maybe related to accessing credit and other production inputs. It may also be about strengthening the voice of farmer, particularly the poor. However the biggest change concerns helping farmers maintain and build profitability in increasingly competitive markets. Another feature, which now seems to be fairly universally acknowledged is that extension will need to involve greater participation of a wider set of stakeholder stakeholders including the private sector. There maybe be ways of privatising some public extension functions, but these may be limited to high value niche crops and animals. Publicly provided services will probably need to concentrate efforts on the marginal and inaccessible areas and for socially disadvantaged groups of farmers and rural households.

A new vision of extension as a locally defined set of approaches that evolve and adapt to meet changing circumstances, suggest that having a national policy is not enough. The capacities of the state and district officials to innovate new and appropriate institutional arrangements need to be enhanced. The case of Indonesia amply illustrated this point.

Policy process challenges

Emerging from much recent experience is the need to move away from a prescriptive top down policy process to one which is more interactive and that iterates between policy and implementation experience. In fact conceiving extension as a set of approaches and tasks which are to a large extent lcally defined, suggest that not only does the policy process need to be quite different, but that also the role of policy also needs to change. For example at one extreme is the case of India with its centrally devised initiatives that can lock extension into long-cycle extension paradigms. At the other extreme seems to be the case of China where extension innovations emerge continuously from the extension practice, with extension policy provide an enabling environment. This is not to say that extension innovations in countries similar to India don't occur – they certainly do. The difference is the ability of the policy process to respond to these innovations and the understanding that the role of policy is to enable rather than only prescribe.

Ways forward

For many agricultural extension systems it is all too easy to define the broad contours of institutional change needed to reform agricultural extension to meet the changing demands placed on it. The current prescriptions include: decentralisation, pluralism, privatisation, cost recovery and so forth. The history and recent developments in Asia illustrates that reform processes only informed by these prescriptions is entirely misconceived and doomed to fail. Now as never before the message for extension in Asia it that the process of reform must be lead from within. And it must be driven by learning about what works and what doesn't and by the nature of local circumstances and context. An analogous initiative in the CGIAR known as the institutional learning and change initiative is trying to do precisely the same thing for agricultural research. These type of approaches stems from the realisation that improving the performance and capacity of a system concerns reflection, learning and incremental change. If extension policy is to pursue such an approach what practical steps could countries take.

A first step would be to undertake an institutional analysis of historical and current experiences of implementing different extension approaches. This should focus on successes and failure and should be undertaken in a constructive manner to devise ways by which these approaches could be modified, bottlenecks removed and institutional arrangements amended. It maybe surprising that currently there are very limited studies and analysis of the extension sector, and these are usually not used in extension policy development and planning. Once again this approach require capacity development as local expertise for analysing complex systems such as extension is lacking at the country and sub-country level (even in India). Without this sort of capacity development countries will remain dependent on international experts to suggest country strategies, models and blue prints.

The next step is to set up extension pilot schemes as experiments in extension. While this in itself is nothing new, such experiments coupled to local capacities in institutional analysis could be used to start and draw broad principles for promoting innovation in rural areas. The contrasts with the conventional approach of establishing pilot extension schemes with a view to refining a new model prior to replication across regions and countries. The final point is that these learning based approaches to reinvent extension need to battle for legitimacy in often highly conservative organisational cultures. Changing these cultures may yet be the biggest challenge of all for reforming extension.

References

ADB (2000). A Study of Rural Asia: Beyond the Green Revolution. (Asian Development Bank) http://adb.org./Documents/Books/Rural Asia/default/asp.

Alex, G., Zijp, W and Byerelee, D. (2002). Rural Extension and Advisory Services - New Directions. Rural Development *Strategy Background Paper No.9, Agriculture and Rural Development Department*, The World Bank, Washington, D.C.

Contado T.E. (1997). Formulating Extension Policy. In: Swanson, B.E, Bentz, R.P.and Sofranko, A.J (eds) *Improving Agricultural Extension- A Reference Manual.* pp.107-114. FAO, Rome.

DAC (2000). *Policy Framework forAgricultural Extension* (draft), Extension Division, Department of Agriculture and Co-operation, Ministry of Agriculture, Government of India.

DAE, (1999). *Agricultural Extension Manual*, Department of Agricultural Extension, Dhaka, Ministry of Agriculture. Government of Peoples Republic of Bangladesh.

Farrington, J., Christoplus, I, Kidd, A.D and Beckman, M (2002). Extension, Poverty and Vulnerbility: The Scope for Policy Reform (Final Report of a study for the Neuchatel Initiative), *Working Paper, 155, Overseas Development Institute,* London.

Gharehyazi, B., Aghajani, M. and Ranjbar,G.A. (2003). Country Paper: Islamic Republic of Iran. In Sharma, R.(ed) Integration of Agricultural Research and Extension, Report of the APO Study Meeting on Integration of Agricultural Research and Extension, Philippines, 18–22 March 2002 (Asian Productivity Organization, Tokyo). http://www.apo-tokyo.org/00e-books/11.AgriResearchExt/12. Ghareyazie.AgResExt.pdf

Gibbons M., Limoges C, Nowotny, H..Schartzman S., Scott P, and Troww, M.(1994). *The new production of knowledge*. Sage, London, pp.192.

Heidary A (2001). Country Paper: Islamic Republic of Iran. In 'Effective Decentralization for Community Development' pp. 176-182 (Asian Productivity Organization, Tokyo). www.apo-tokyo.org/icd/papers/E-Publications/01.EffDecforICD/3-06.pdf

Kadir,I et.al. (2003). Country Paper: Indonesia. In Sharma, R.(ed) Integration of Agricultural Research and Extension, Report of the APO Study Meeting on Integration of Agricultural Research and Extension, Philippines, 18–22 March 2002 (Asian Productivity Organization, Tokyo). http://www.apo-tokyo.org/00e-books/11.AgriResearchExt/11.Kadir.AgResExt.pdf.

Katz, E (2002). *Innovative Approaches to financing extension for agricultural and natural resource management- Conceptual considerations and analysis of experience*. Swiss Centre of Agricultural Extension, Lindau.

Malik, W (2003). Operationalizing Agricultural Extension Reforms in South Asia: A Case of Pakistan. Paper presented at the Regional Workshop on 'Operationalizing Reforms in Agricultural Extension in South Asia' New Delhi, India, May 6-8, 2003. (The World Bank Group Washington, D.C.) http://www.worldbank.org/wbi/sdruralpoverty/ag_extension1/Materials/Plenary/ Pakistan Country PaperonAgExtension Reformsfinal.pdf

Neuchatel Group (2002). *Common framework on Financing Agricultural and Rural Extension, Neuchatel Group.* Swiss Centre for Agricultural Extension and Rural Development, Lindau.

Qamar, MK (2002). Global trends in agricultural extension: challenges facing Asia and the Pacific region. SD Dimensions: Knowledge : September 2002. FAO, Rome. http://www.fao.org/sd/2002/KN0903a_en.htm

Rivera, W.M., Qamar, M.K. and Crowder, L.V. (2001). Agricultural and Rural Extension Worldwide: Options for Institutional Reform in the Developing Countries. Extension, Education and Communication Service, Research, Extension and Training

Division, Sustainable Development Department. FAO, Rome. http://www.comminit.com/ redirect. cgi?r= ftp://ftp.fao.org/docrep/fao/004/y2709e/y2709e.pdf

Sharma, R.(ed) (2002). Integration of Agricultural Research and Extension, Report of the APO Study Meeting on Integration of Agricultural Research and Extension, Philippines, (Asian Productivity Organization, Tokyo). http://www.apo-tokyo.org/00e books/11.AgriResearchExt.htm

Sulaiman, V.R. (2003). Innovations in Agricultural Extension in India. SD Dimensions: Knowledge: FAO, Rome. http://www.fao.org/sd/2003/KN0603_en.htm.

Sulaiman, V.R. and Hall, A.J. (2002). Beyond Technology Dissemination: Reinventing agricultural extension, *Outlook on Agriculture*, Vol 31, No 4, pp 225-233.

Sulaiman, V.R. and Hall, A.J. (2003). India: The emergence of Extension-Plus: Future for Extension Beyond technology transfer? In 'Rivera, WM and Gary Alex (eds.) Extension and Rural Development'. Washington D.C., The World Bank.

Suryana, A and Erwidodo (1996). Agricultural Policy Reforms In Indonesia: Accelerating Growth With Equity, (Center for Agor-Socioeconomic Research Agency for Agricultural Research and Development, Indonesia). http://www.agnet.org/library/article/eb434.html

Swanson, B (Ed) (1990). *Global Consultation on Agricultural Extension: A Report.* FAO, Rome.

Uddin,R (2003). Existing Extension System: Strengths, Weaknesses and Proposed Reforms in Bangladesh, Paper presented at the Regional Workshop on Operationalizing Reforms in Agricultural Extension in South Asia New Delhi, India, May 6-8, 2003. Washington D.C., The World Bank Group. http:// www.worldbank.org /wbi/sdruralpoverty /ag_extension1/Materials / Plenary/ Bangladesh Country Paper. pdf

Yonggong. L. (1998). Institutional and Policy Reforms or Rural Extension in China during the transition towards a market economy, *Training for Agricultural Development, 1997-98,* FAO, Rome.

Zakaria, A. (2003). Decentralizing Extension to Local Governments: Indonesia Experience. Paper presented at the Regional Workshop on Operationalizing Reforms in Agricultural Extension in South Asia New Delhi, India, (The World Bank Group).

3

Changing Roles of Agricultural Extension for Socio-economic Development in the Republic of Korea

Cho, Yeong – Cheol & Yang, Seung – Choon

Introduction

Recently, there has been rapid change in agricultural extension in Korea with progress in globalization, localization, and knowledge-based society. Notable was the Korean agricultural extension's devolution from national to local government in 1997.

At that time when agricultural extension workers served for the national government, they disseminated information on matured technologies of the research institutes of the Rural Development Administration (RDA) through local extension institutes nationwide. After the conversion into local government, however, agricultural extension workers have performed various projects fit to local situations alone.

Since the devolution, Korea's agricultural administration has been patterned after the open-door agricultural administration system under the World Trade Organization (WTO). Today the country is facing some problems in persuading Korean farmers questioning the National Assembly ratification's result of negotiation on rice this year. In addition, it is inevitable for Korea to turn its way in agriculture especially with the advent of the Doha Development Agenda (DDA).

Meanwhile, although agricultural production has increased over the years,, consumption has decreased. At the same time, instability of the farms' income is getting bigger with the price of farm produce getting lower. Creating new workforce in the rural areas is difficult because rural population decreases and structure of population in Korea changes. The number of the people aged over 65 is rapidly increasing. For these reasons, Korean agriculture needs to take action against this severe challenges.

In order to overcome the situation, Korea has established comprehensive plans for farming and rural communities and tried to respond to the purpose of agricultural extension services. Medium- and long-term projects have also been implemented.

Agricultural Changes in Korea

Table 1 shows the rapid changes in Korea's economy since 1960. Korea was one of the poorest countries in the world with US$79 per capita gross national income (GNI) in the 1960s, but only 40 years later the country's GNI reached US$10,000.

Table 1. Changes in some Domestic Economic indicators in Korea

Class	1960	1970	1980	1990	2000	2004
Value added in agriculture as % of GDP		29	16	09	05	04
Farm population as % of total	28	45	28	16	09	07
Farm household income in won US$ 1 is approximately 1050 Korean won			256	2693	11026	23072

In the same period the portion of national income that came from agriculture/forestry/fisheries decreased steadily from 29.2% in 1970 to 5% in 2000, the number of farm households as % of all households dropped from 53.7% in 1960 to 8% in 2004, farm population also dropped from 58.3% in 1960 to 7.1% in 2004. As a result, cultivated land per farm household increased to 1.48 ha in 2004, but farms are still small.

Establishment of RDA and Evolution of Agricultural Extension Services

A. Agricultural extension organization before the establishment of RDA

After the Second World War, the first agricultural extension organization in Korea, the Institute of Agricultural Improvement, was established in 1945. This organization was operated following the American way of consolidating colleges of agriculture, agricultural experiment stations, and the extension service bureau. Nevertheless, the organization did not progress because of lack of experts, budget, and understanding of the organization and doctrine of agricultural extension in the United States. Nevertheless, the Institute of Agricultural Improvement was significant in shifting from the mandatory and unilateral agricultural extension during the Japanese occupation to the democratic and educational agricultural extension of the United States.

With the establishment of the Korean government in 1948, the Institute of Agricultural Technology was founded by consolidating agricultural experiment stations and extension bureaus, after transferring colleges of agriculture to the Ministry of Education, but this operation stopped during the Korean War (1950–1953).

Established in 1957, the Institute of Agriculture commenced the farm household guidance project by law, set up an administration system, allotted the necessary budget by the assembly action, and arranged professionals with sufficient knowledge and skills to work devotedly for farmers.

B. Inauguration of RDA and the agricultural extension campaign

Real agricultural extension services in Korea started in 1962 after the proclamation of the Rural Development Law. This Law prescribed comprehensive range of extension programs so all kinds of education under rural development have been called "agricultural extension" from then on.

The Law also stated that it was the task of the Rural Development Administration to conduct agricultural extension. Other organizations could only do this with the approval of and in cooperation with the RDA.

As shown in Figure 1, RDA was established as the extension and research arm of the Ministry of

Agriculture and Forestry. Under RDA, there were research and experimentation stations for development of new varieties and production techniques.

In each province, there was a Provincial Rural Development Administration under the administrative control of the provincial governor. At the local level (county or city), there were city/county extension offices. Under the jurisdiction of the city/county extension offices, grassroots extension offices were placed in every two to three towns to conduct extension projects directly at the farms.

The grassroots extension offices were set up in every town to conduct extension services aimed at achieving food self-sufficiency in 1975. This grassroots extension system lasted until consolidated into city/county extension offices in 1989.

Figure 1 The Rural Development Administration

Ministry of Agriculture and Forestry
- Rural Development Administration
  - Research Bureau
  - Extension Bureau
  - Training Institute

Provincial government
- Provincial Rural Development Administration

Head of City/County
- General Affairs section
- Extension Section
- Research Section
  - City/County Agricultural Extension Office

Head of Town
- Grassroots Agricultural Extension Office

Changes and Characteristics of Agricultural Extension in different periods

A. Agricultural extension during the period 1962 - 1975

1) The military government started in 1961 to implement development strategies centered on the industries to boost export, so the role of agriculture and rural areas complemented the industries and cities.
2) At first, food self-sufficiency was the goal of agricultural policies but this was changed to sustainable supply of cheap farm products for production and export of cheap industrial products.
3) Lack of workforce in rural areas was not a serious problem.
4) Farmland area per farm household was 0.9 hectare. Farmers couldn't find an other job. Their farming system was based on irrigated rice. Their income from non-agricultural activities was relatively low.
5) The farm size was not adjusted to the introduction of farm machinery and the level of consumption of fertilizer and agrochemicals stayed low.
6) The demands of farmers reflected on farmer movements showed that farmers thought the problems in agriculture were related to lack of technologies and of cooperation among themselves. The farmers couldn't understand the connection of agricultural problems with social structure problems.
7) In this period 57% of the rice was imported and 9% of the horticultural and cash crops consumed. The food self-sufficiency was 90%.

B. Agricultural extension during the period 1975 -'89.

From 1975, when agricultural extension organizations in every town were established, to 1989, the period of abolition of the grassroots extension offices, the situation in the agricultural extension are as follows:

1) Feminization of agriculture and a shortage of farming successors caused by industrialization.
2) While cultivation of food crops decreased, cultivated land for vegetables, specialty crops, fruits, and livestock increased and the number of farmers planting specialty crops increased.
3) On farm household income from non-agricultural activities rose.
4) There was a drop in consumption of locally produced rice and barley as a result of growing import of farm products, changes in food consumption, and increased consumption of wheat.
5) The diffusion rate of farm machinery started to rise and the amount of consumption of agrochemicals and fertilizers also increased gradually.
6) The farmer organizations emphasized the demand on cooperative projects and democratization of agricultural cooperatives. In this period, many started to think that the cause of problems in agriculture was not attributable to the lack of agricultural technology, butto social structure.
7) In this period 52% of the rice was imported and 26% of the horticultural and cash crops and 14% of the livestock products consumed. The food self-sufficiency was 40%.

C. Agricultural extension after 1989

With the integration of the grassroots extension offices into the city/county extension offices in 1989, agricultural extension services were reorganized and became specialized. This period shows and the following features:

1) With the progress of democratization and localization, freedom was highly celebrated.
2) Agricultural policy aimed at improving international competitiveness of farmers in growing cash crops to cope with the opening of markets for farm products.
3) Although average farm size was only 1.2 hectares, the area for horticultural facilities was expanded and the interest of farmers in cash crops got higher.
4) The ratio of farm population to total population decreased to 13%. Particularly, with the youth and the middle-aged leaving the farmland, feminization and graying of farm workers was intensified. Meanwhile, the shortage of farm workers led to expansion of uncultivated land reaching 3% of the total rural area.
5) The degree of food self-sufficiency plunged to 34% and grain consumption per capita gradually decreased, while consumption of vegetables and fruits increased little by little.
6) The shortage of farm workers also boosted the rate of agricultural mechanization. Mechanization in planting irrigated rice reached 90% and this meant that almost all operations were mechanized.
7) The demands from farmer organizations notably changed to political matters such as opposition against importation of farm products and a plea for democratic unification, as well as economic matters and various activities.

8) In this period 41% of the rice was imported and 33% of horticultural and cash crops and 20% of the livestock products consumed. The food self sufficiency was 34%.

The Rural Development Law and Changes in Agricultural Extension Services

The Rural Development Law of 1962 states that the objectives of RDA are to contribute to the development of farmers' welfare through conducting agricultural experiments and research, transfer of scientific techniques and knowledge sharing on agriculture and rural life, and training of rural leaders and farmers.

The law stipulated to conduct diverse projects such as improvement in farming and rural life as well as development of local society and training of rural youth and to unify rural extension system by preventing the national or local governments in conducting rural extension services without complying with the law.

Agricultural extension services have been modified with evolving various activities since 1962. The characteristics of changes in certain periods are shown in the following.

A. Characteristics of agricultural extension services in the period 1962 - 1974.

Agricultural extension services from 1962 to 1974 had a real grassroots extension office under the jurisdiction of the county extension office. The extension educators were stimulated to conduct extension services by visiting farm households and instructing farmers through the local allotment system. The characteristics are as follows:

1) The ultimate goal of agricultural extension services is the improvement in farmers' welfare. The agricultural extension services' operating objective was to complete the Green Revolution, a national goal, which meant food security through an increased yield of rice.
2) The agricultural extension system of organizations was established as a four-stage perpendicular system composed of central, provincial, city/county, and towns organizations. Real grassroots extension organizations, the first-line organizations for agricultural extension, took charge of two to three towns. City/county extension organizations were organized in the '60s with extension, development, and technique sectors, so they were more organizations for community development. And, crops, livestock, and cash crop sections were created. The organizations for community development and extension of farming techniques were about equal in size.
3) The quality level of agricultural extension services can be assessed by the academic career of extension educators. At that time, most extension educators were college graduates. Meanwhile, food crops occupied 20% of budget allotment for agricultural extension services.
4) Extension of techniques for agricultural production and improvement of rural life, training of rural youth students, community development, and other activities related to agricultural development were widely conducted. Because a new high-yielding variety, Tongil rice, was developed and disseminated, the efforts to increase food production were reinforced.

B. Characteristics of agricultural extension services in the period 1975 -1988.

By establishing real grassroots extension organizations in 1975, agricultural extension services expanded. These real grassroots extension organizations served as central organizations until 1988. The characteristics of this period are as follows:

1) The ultimate goal of extension services, the improvement of farmers' welfare, was not changed, but the operating objective was changed into completion of the Green Revolution. Self-sufficiency in rice became the key operating objective.
2) The extension organization system established in the early 1970s was maintained and two to three agricultural extension educators at a real grassroots extension organization conducted services such as transfer of agricultural production techniques and development of rural communities.
3) In the academic background of agricultural extension educators, the number of college graduates, 31% of the entire extension educators, was 10% lower compared with that of the previous period because extension educators were appointed mainly to stimulate the cultivation ofTongil rice.
4) The agricultural extension services of this period contributed greatly to stable supply of staple grains, thereby achieving the goal by concentrating on dissemination of new high-yielding variety. With the improvement in distribution of new high-yielding variety, the use of agrochemicals and fertilizers consequently increased. The World Bank estimates that in Korea in 1999-2001 the average yield of cereals was 6500 kg/ha, 30% more than 20 years earlier. Despite the completion of the Green Revolution, agricultural extension for fostering

farmer organizations, which were the key method in solving farmers' problems, was insufficient.

C. Characteristics of agricultural extension services

Between 1975 and 1989 when agricultural extension services were conducted with a real grassroots extension organization as center, agricultural extension rapidly changed and the conversion of agricultural extension services was accelerated. By integrating the real grassroots extension organizations into city/county extension organizations in 1987, there was an attempt to establish a specialized professional extension system on certain crops. Nevertheless, by establishing the farmers' counseling offices at the previous real grassroots extension organizations in April 1991, two years after the withdrawal of these organizations, the agricultural extension educators were stationed at the offices to counsel about agricultural development and 1,467 real grassroots extension organizations were reorganized under city/county extension organizations. The characteristics of this period are as follows:

1) The operating objective of agricultural extension services was to promote the welfare of rural communities through on production techniques to increase crop yields and improve the quality rural life. Concrete extension programs, however, were insufficient.
2) Of the agricultural extension educators 39% were college graduates a bit more than in the previous period, but this could not be considered sufficient in conducting specialized extension.
3) The system of city/county organizations was specialized into three divisions—extension division, social development division, and technique

dissemination division. Nevertheless, specialized extension programs, because the division received insufficient technical support.

4) The contents of extension by professional items shows that in food crops, the extension for production increase was changed in extension for stable production of good quality rice. The extension services for cash crops, such as horticultural and flower production, were reinforced to help growers to compete with foreign growers. Limited availability of personnel, budget, and a specialized organization, however, were insufficient to realize this goal.

The characteristics of agricultural extension services after 1962 are shown in Table 2.

Agricultural extension services anywhere in the world are now facing a transition period. This holds true in Korea also.

RDA was established to conduct projects prescribed in the Rural Development Law of 1962, and written in the comprehensive report on the improvement of agricultural extension system in 1965. Discussions about the area, goal, and leading organization of agricultural extension services continued, and there have been changes in the organization system.

The agricultural extension services are conducted through organizations and these organizations can develop or dwindle through interacting with other organizations.

Analyzing which circumstances influence the agricultural extension organizations in which way is most crucial in establishing the direction for development of agricultural extension services. Korea's

Table 2. Change of characteristic of agricultural extension services

Classification	1962–1974	1975–1988	1989–1996
Organizational Characteristic			
Ultimate Goal	Promotion of farmers' welfare	Promotion of farmers' welfare	Promotion of farmers' welfare
Operating Objective	Completion of the basis for the Green Revolution	Completion of the Green Revolution	Achievement of rural communities with improved welfare
Organizational System	An extension organization in every two to three towns	An extension organization in every town	Consolidation of real grassroots extension organization
Extension Workforce	High-level academic background	Low-level academic background	Improved level in academic background
Major Field of Extension	Food crops	Food crops	Cash crops
Budget for Extension	National expenditure mainly	Increase in local expenditure	Local expenditure mainly
Characteristic of Each Field			
Technologies on Rice	Early planting of irrigated rice	Cultivation of a new high-yielding variety	Stable production of high-quality rice
	Collective farming of aquatic rice		Mechanized transplantation ofyoung rice seedlings

Technologies on Dry-land farming	Potatoes, beans, maize, a variety of barley, sweet potatoes	Potatoes, beans, maize, a variety of barley, sweet potatoes	Potatoes, beans, maize, a variety of barley, sweet potatoes
Technologies on Crop Protection	Forecasting of crop diseases and pests	Forecasting of crop diseases and pests	Forecasting of crop diseases and pests
		Publication of situation about diseases and pests	Publication of situation about diseases and pests
Technologies on Livestock	Farming superior feed crops	Farming superior feed crops	Farming superior feed crops
	Breeding technologies in livestock	Breeding technologies in livestock, prevention of diseases	Breeding technologies in livestock, prevention of diseases
Farm Management	Design for diagnosis of farm management	Extension on farm management and marketing of farm products	Specialized extension on farm management
Development of Communities	Comprehensive development of rural communities	Comprehensive development of rural communities	Development of local agriculture
Improvement of Rural Life	Rationalization in rural life	Improvement in food, clothing, and housing	Improvement in the quality of rural life
Training of Rural Youth	Fostering of farming successors (mainly farmers)	Fostering of farming successors (farmers + students)	Training of democratic Citizens (mainly students)

agricultural extension experienced various drastic changes after 1962. The results of the analysis on changes in characteristic of agricultural extension services are as follows:

The circumstances of the period 1962 -1975 established food self-sufficiency as a priority in social economy and agricultural policies. Workforce in farming was sufficient and farm families could earn by planting food crops. With regard to the demands of farmers shown in the farmers' movement, problems in agriculture were attributed to a lack of agricultural technologies. The agricultural extension services of this period gave more emphasis on disseminating technologies for farm production rather than the development of rural communities. Thus, food self-sufficiency was achieved through distribution of a new high-yield rice variety. During this period, agricultural extension services fit to farmers' needs were conducted.

Rural Extension Services in Korea after Decentralization

A. System of agricultural technology development and diffusion

Most of the agricultural technology innovation system in Korea is controlled by the public sector. This system is composed of diverse central and local government agencies. RDA has eight research institutes (National Institute of Agricultural Science and Technology, National Institute of Crop Science, National Institute of Agricultural Biotechnology, National Institute of Agricultural Engineering, National Horticultural Research Institute, National Livestock Research Institute, National Institute of Highland Agriculture, National Institute of Subtropical Agriculture) and an agricultural college. The staff at these central government offices totals 2,087 (1,166 researchers, 71 extension specialists, 317 administrative officials, and 533 technicians). New agricultural technologies, which were developed in RDA, are delivered to extension officials working at the Provincial Rural Development Administrations (PRDA) or the City/County Extension Centers (CEC). These extension specialists transfer new agricultural technologies for farmers' adoption.

All nine provinces have a PRDA, which is composed of an extension board and a research board. All over the country, there are 35 Specialized Crops Research Institutes (SCRI), which are controlled by PRDA. About 10 persons work in each SCRI. Each SCRI focuses on one crop only.

B. Change in agricultural extension organization and manpower

Since the decentralization in 1997, most extension specialists have experienced a change of status from central government officials to that of local government officials. This means that the personnel management policy has been matched to the system of agricultural extension organization. With the tendency of decentralization, agricultural extension services have been developed according to the situations of local areas. After the decentralization, the number of CECs has increased. In 2004, a CEC was established in Ansan City, where there was no CEC before the decentralization. This implies that the local residents recognized the importance of agricultural extension services. After the decentralization, however, workforce dynamics has been affected. For 38 local governments, the agricultural extension services and agricultural administration services have been integrated so that the quality of agricultural extension service has gone low.

There were 1,428 farmers' counseling offices (FCO) before the decentralization. These were the offices that quickly provide agricultural technology to farmers and were controlled by the local CEC. After the decentralization, the number of FCOs has been radically reduced to 570. Recently, the number increased to 625.

The number of agricultural extension officials decreased to 4,906 or one per 280 farm households. In many other countries are more farm households per extension official. Before the decentralization, 6,842 extension officials worked in agricultural extension services all over the country. This reduction of agricultural extension services caused some problems in the transfer of agricultural technology to local farmers. To cope with this lack of extension workers, the agricultural extension services have developed diverse extension methods to which information technology were applied. It is expected that the number of extension officials will not be changed.

Table 3. Change of extension organizations and personnel after the decentralization

	Before Decentralization (1996)	After Decentralization (1998)	Present (2005)
Extension Organizations			
- RDA	1	1	1
- Provincial RDA	9	9	9
- City/County extension center	162	157	160
- Farmers' counseling office*	1,428	570	625
Extension Officials (Total)	6,842	5,545	4,906
- RDA	94	85	71
- Provincial RDA	289	241	241
- City/County extension center	6,459	5,219	4,594

* One extension official works at every FCO.

C. Important missions of rural extension services in Korea

With the Green Revolution, the agricultural extension services contributed to Korea's self-sufficiency in rice production. Korean agriculture and rural communities, however, experienced a crisis brought by the increase of imports of agricultural products, which has been accelerated by DDA, Free Trade Agreement, among other things. On this background, RDA created a master plan to improve the quality of Korean rice to help rice farmers respond to the radical change of the rice market. Since 1970, Korean economy has been restructured on manufacture and service industry. Farming has become such a difficult job so that many farm households gave up farming and left the rural communities. At present RDA is making a counter-plan to revitalize the rural communities. While there has been strong opinion that the range of agricultural extension services' clients should be expanded to consumers, diverse extension services were developed and exercised. The main missions of agricultural extension services in Korea are the following:

1) Improved rice quality

In 2004, Korea and some exporting countries agreed to increase obligatory imports of rice by 408,000 tons until 2014. Obligatory rice imports in 2005 accounted for 225,000 tons. Faced with this change in rice market environment, RDA has planned and implemented projects that aimed to facilitate the production of high-quality rice. For example, the "Top Rice" project aims to help rice farmers in increasing the proportion of good polished rice in the packages, which consumers buy, to reduce the use of nitrogen fertilizers and to prevent mixing diverse varieties of rice into one package. In every CEC, an Office of High Quality Rice has been set up. At the same time, specialists are assigned to identify the difference among rice varieties and to manage the quality of rice. The 19 "Top Rice"-producing demonstration areas were designated, in which 1,500 farmers from all over the country participated. The total size of said demonstration areas is 1,660 hectares. The "Five-year Plan in Breeding High-quality Rice" was implemented in RDA. In addition, RDA launched a pilot project on seed proliferation, which focuses on quick dispersal of newly developed rice varieties. RDA also systematized diverse technologies, which are helpful in reducing the production cost. Though RDA has disseminated standardized rice cultivation techniques, diverse rice cultivation manuals for different areas are yet to be developed and disseminated in the near future.

2) Facilitating environment-friendly farming practices

The importance of environment-friendly farming is recognized more and more. Agricultural extension services in Korea have set up and operated some Integrated Environment-friendly Farming Areas. The purpose of this pilot project is to establish the system of agricultural techniques, which can be used for environment-friendly farming. For the past decades, RDA has implemented projects focusing on rice cultivation. In the future, however, RDA will move its project focus toward establishing viable rural communities, including village environment, rural amenities, and green tourism among other things. Monitoring farming practices will be strengthened to conserve agro-environment and to produce safe agricultural products. Public relations for consumers and farmers should be reinforced so that people will recognize the importance of environment-friendly agriculture. With these projects on environment-friendly agriculture, the system for Integrated Pest Management and quick measures against weather disasters are maintained.

3) Diverse projects to revitalize rural communities

Rural communities in Korea are faced with an aging population. RDA is conducting a project that will establish 800 "rural health and longevity villages" all over the country by 2009. In 2005, 100 "rural health and longevity villages" were made and are being operated. In these villages, appropriate jobs are provided to the aged. Also, extension workers are helping the aged in improving their health and diet and in facilitating their participation in adult learning and volunteer programs. RDA strives to improve the living environment of the "rural health and longevity villages."

RDA has initiated the promotion of green tourism. The project Rural Traditional Theme Village was launched in 2002. This project aimed to make one rural traditional theme village in every county in Korea. The residents in those villages will develop their own communities so that they will be able to provide visitors from urban areas with invaluable information on rural communities.

Moreover, additional pilot projects, aimed at helping farmers to manage their farmland safely, were launched. Various policies for women farmers are also implemented. Particularly, in helping rural communities use their own resources for endogenous development, RDA launched the pilot project "Korean Traditional Dietary Culture." For urban citizens, information is provided in planning their rural lives after retirement.

The conceptual frameworks of these projects and plans are different from the past agricultural extension programs. These are pilot projects aimed at sustainable development of rural communities.

4) Improved quality of fruits

Quality of fruits became more important than ever. There is no standard quality, however, to which farmers refer in their production process. Therefore, RDA designed some projects that would set a quality standard for apples and pears to improve the quality of fruits and reduce the production costs. These projects will be implemented by the Quality Management Association of Fruits, which will be organized based on the partnership of RDA with farmer organizations.

5) Measures to prevent animal epidemics and to recycle livestock excrements

Disease control is an important concern in livestock farming. Globalization forced livestock farmers in Korea to give attention to livestock diseases from other countries. Thus, systematizing the prevention of livestock diseases is a very important mission. Also, recycling livestock excrements became an important issue in Korea. RDA is prioritizing the development and dissemination of technique in converting livestock excrements to compost and liquid manure.

6) Training specialized farmers

The main goals of agricultural extension services are transferring agricultural technologies and increasing the competence of bringing up specialized farmers. It is quite difficult, however, to ensure that specialized farmers live in their own rural communities. The number of empty villages, where a baby's cry cannot be heard, are increasing in Korea. To increase human capital in the rural areas in Korea, young farmers should be able to sustain their job in the rural communities in the future. RDA is planning to help farmers build learning organizations for themselves and to support those organizations so that the farmers will be capable of producing, processing, and selling their products profitably. Also, by strengthening support for the students in the Korean National Agricultural College, RDA will assist them in settling in rural communities as successful farmers.

7) Re-establishing relationship with local agricultural extension organizations

As mentioned earlier, after the decentralization in 1997, agricultural extension services have been restructured so that projects and programs, which the local governments need, can be selected and supported appropriately by the central government. Nevertheless, RDA still has the responsibility in developing and diffusing new agricultural technologies. Also, RDA confirms the budgets for those projects. After eight years, we should find the solution by planning and implementing good projects for farmers and consumers based on mutual cooperation between the central and the local governments. The central government should not only provide the budget that local governments need but also implement various projects concerning balanced development across the nation. The local government should clarify their own roles and reinforce cooperation network with other institutes because more budget, technologies, and information are needed.

Conclusion

In Korea, modern agricultural extension services started with the establishment of RDA in 1962. At that time, RDA, PRDA, and CEC were organized. Though the provincial and the city/county's organizations belonged to the local government, they were controlled by the RDA. The administrator of RDA had the personnel commission of local extension specialists until 1996. Therefore, the autonomous projects of local extension organizations were somewhat limited. After 1997, however, the personnel commission of local extension specialists was transferred to presidents of local governments. The central government's control over the agricultural extension services weakened; the bottom-up approach of agricultural extension was furthered although this does not mean that the policy, budget, and technical support of RDA are not necessary.

In the past, the roles of rural extension organizations were to help farmers increase their productivity, improve their living environments, and train farmers. In doing so, rural extension services achieved the Green Revolution and the White Revolution (=milk) in the 1980s. There was no change in the objectives that rural extension should transfer agricultural technology and educate specialized farmers. But after the decentralization, the roles were more diversified to solve various problems in rural communities. In the future, diverse projects that are in accordance with the trends of globalization and localization should be designed and implemented in the rural extension services in Korea.

References

Cho, Yeong-Cheol (2000). Agricultural Technology Transfer and Training Program in Korea. Paper presented at the *First Workshop on*

Agricultural Technology Transfer and Training, Agency for Agricultural Research and Development, Ministry of Agriculture, Jakarta, Indonesia.

Cho, Yeong-Cheol (2003). *Agricultural Extension Program in Korea, Agricultural Technology Transfer and Its Consequences.* Agricultural Research Institute, Council of Agriculture, Taiwan.

Rasheed Sulaiman V, Andy Hall (2004). Towards Extension-plus: Opportunities and Challenges. National Centre for Agricultural Economics and Policy Research, New Delhi, India.

Rivera, W, M & Gustafson, D. J. (1991). *Agricultural Extension: Worldwide Institutional Evolution and Forces for Change.* Elsevier Science Publishers, Amsterdam, Netherlands.

Rivera, W. M. et al. (2001). *Agricultural and Rural Extension Worldwide: Options for Institutional Reform in the Developing Countries.* FAO, Rome.

Van den Ban, A.W. (2004). World Trend in Agricultural Extension. Paper presented at the *Second International Conference on Agriculture Education and Environment, Suwon, Korea.*

World Bank (1990). *Agricultural Extension: The Next Step.* Washington D. C: World Bank.

Yang, Seung-Choon (1994). *A Study on the Transformation of Korean Agricultural Extension Services Associated with Their Organizational Environment.* Ph. D. dissertation, Seoul National University, Suwon, Korea.

Swanson, B. E., (1997). *Improving Agricultural Extension.* Rome: FAO.

Swanson, B. E. (2004). "Extension Strategies for Poverty Alleviation in a Global Economy." Paper presented at the *Second International Conference on Agriculture Education and Environment,* Suwon, Korea.

4

CHANGING TASKS OF EXTENSION EDUCATION IN INDIAN AGRICULTURE

R. Dwarakinath

Agricultural extension is an instrument of change.

It enables farmers to seek and adopt desirable changes in farming.

Its focus is on technology improvements, leading to human resource development.

In every culture, agriculture normally keeps changing, either slowly or rapidly.

The role of agricultural extension is to make this process more purposeful.

As the economy expands and technology advances, agricultural extension does not remain a matter of choice, but becomes a socio-economic necessity.

And, as the agricultural sector changes, the tasks of the agricultural extension

will also change, while its basic functions remain essentially the same.

Present chapter is based on a specific premise. According to this position, agricultural extension, as one particular system of extension education, is essentially a process of intervention. By definition, the extension system is external to the rural society, and strives to provide access to useful and practical knowledge, relevant to farmers' problems and opportunities. Sometimes, it represents the out-reach efforts of a prominent knowledge centre in a developing society. But, more often, it forms a major part of the integrated rural development undertakings of the country. Normally, in relating new knowledge from external sources to local needs, the extension system diligently builds upon the prevailing traditional knowledge systems. By this means, on one hand, it widens the horizon of the farmers, and, on the other, it enhances the quality of local farming practices.

In India, over the years, extension system has come to be perceived in different ways. Very often, the development departments of the government look upon this as a delivery mechanism in the farm sector. Then, again, in a paternalistic stance, it is seen by some public agencies as a benevolent system that bestows some betterment opportunities in farm life. Very rarely, however, it is recognized as a system that enables farmers in improving their own perceptions, attitudes and abilities, in their ways of living, and in making a living. It is this last perception that pervades this paper.

The attempt in the paper is to provide a simple picture of the village community as the social arena where development events take place, and the agricultural extension system operating as a change agency. It deals with a brief account of the changes taking place in the agriculture sector, and their development implications. It then devotes attention to some of the prominent extension tasks of contemporary relevance.

A. Perspective

Generally, agricultural extension system is put in place to work with farmers in the village communities. These communities are social systems having their own culture, norms and traditions. They also have past experience that pervades their present. All farmers, as individuals and groups, are conditioned in their behaviour by the cultural norms and past experience. Similarly, their farming systems, too, are influenced by these factors. That is why the changes brought about in agricultural practices are often regarded as cultural changes in essence.

Farming, for the most part, is a way of making a living. In these ways of living and making a living, changes do occur, from time to time, changes in all communities. They may arise due to some changes in the natural farm resource base like access to irrigation, or due to changed needs of the farm families like recourse to market farming, or due to changes in the farm technologies like the advent of organic farming. Such changes may arise internally in the community, or externally by way of emerging new opportunities.

In the particular case of India, the propagated changes in farming are very much a part of the externally introduced development initiatives. They are perceived as induced or guided changes in a traditional farming system. They are usually a part of the larger national reconstruction undertakings. Thus, extension education is really a venture in agricultural modernization, intended to stimulate rapid and extensive changes in the farm sector, seeking livelihood improvements, poverty alleviation and economic progress. The ultimate aim, however, is human resource development.

The focus of extension system is on the farming communities. In almost all the societies, agriculture happens to be an important segment of the economy. Especially in the developing countries, it is both the dominant means of livelihood for the majority of the population, and, at the same time, a substantial contributor to the national economy. Further, in these countries, as they develop in many ways, agriculture generally lags behind the other segments of the economy. Therefore, most of these nations, embarking upon a process of modernization, seek a rapid up-gradation of the agricultural sector. Agricultural development, therefore, becomes a prominent goal in these national regeneration efforts. In this context, agricultural extension becomes a chosen means to an end.

The nature of agriculture, as an occupation, is a well-known matter. Still, it is helpful here to bring into focus its basic features. The foundation for farming is a set of natural resources. There is no farming without this base. Every farmer works with a farm, which is a combination of the natural resources of soil, water and biodiversity, under a given climatic condition. Using this base, he employs the biological life processes of plants and animals, to produce some farm goods. What he does here, basically, is the manipulation of the soil, water and bio diversity, under a climatic situation, ensuring that the chosen plant and animal species go through their life cycles. Thus, in broad terms, farming constitutes the manner in which farmers manipulate the natural resources accessible to them, and manipulate the production processes chosen by them.

Exploitative land-use practices

From a broader perspective, the way in which farming is being practiced all these years is also a matter of concern. We know that human communities are an integral part of the eco system. They are born out of the eco system and live upon the eco system. The ways in which people make use of the natural resources are generally recognized as farming, logging, mining and fishing. In farming, by far the largest human activity, they make use of the natural resources like soil, water and bio diversity, in a given climatic situation, for making a living, and lately for making money as well. Some of these resources are renewable, within limits, and others are not. Then, the manner in which these natural resources are utilized may amount to "exploitation" or "conservation" of these resources. It is necessary, in the larger interest, our resource use pattern in farming is more conservational rather than exploitative.

However, the relentless economic pursuits of the present-day human communities have often tended to be exploitative. The rapid growth in science and technology has been sometimes reinforcing this process. Farming is an extensive human intervention in the eco system. The contemporary farming patterns have been witnessing the exploitation, in all its severity. In this regard, the developing countries with huge populations seem to be the worst sufferers.

There is another aspect of the human intervention with the eco system. In most societies, the sections that are better endowed seem to be gaining a greater access to the available farming resources. As a result, the poorer segments of the population are left with lesser opportunities to make a decent living. This differential access to the natural resources has generated, over time, enormous economic, social and gender inequities in the society. This is manifested more glaringly in the case of developing countries.

India is a major developing country seeking rapid development, after it attained Independence. Therefore, a "leap forward" approach was adopted in the early days of Freedom. Green Revolution, for this purpose, was an excellent option. But, in retrospect, it is observed that this remarkable innovation, with its focus on "crops and yields," has just served the purpose of enhancing the food production, and did not address the other aspects of agricultural development. As such, today, in the farm sector, we have some outstanding achievements along with some disturbing lapses.

From this viewpoint, agricultural extension is essentially a means to improve, over time, the manipulative abilities of the farmers. Also, with the increasing population, changing life styles and access to powerful technologies, farming can easily become a major source of damage to the environment. It is essential, therefore, that the extension system makes special efforts to build the sensitivity and capacity of the farming communities with regard to eco friendly farming alternatives, as opposed to exploitative systems.

Knowledge as a source of growth

At the beginning of the development process in any country, the prevailing system of agriculture will be mainly "traditional," in nature. It will primarily be a part of the culture of the society in the region, in terms of the ways in which environment is dealt with, and the ways in which livelihoods are pursued. The farming practices in the traditional system, at any given time, are the manifestation of the experience and wisdom of the many generations in the past. A major tendency in this system will be "conformity" to the prevailing norms rather than "deviation." Hence, usually, the room for innovations in the traditional system of agriculture will be somewhat limited.

However, in the last two centuries, modernization has become a universal phenomenon. A salient trend in this process is the growth of knowledge centres like universities, scientific institutions and research establishments, and the accumulation of a large body of useful knowledge. Quite often, a good part of this knowledge, in the spheres of physical, biological and social sciences, will have some scope for direct or indirect application in the farming pursuits. Hence, this useful knowledge has to find its way to the far away farms and homes.

But, in the past, there was a formidable problem to be dealt with in this regard. Transmitting the potentially useful knowledge from the centers where it is generated to the isolated farming communities where such knowledge is needed remained a difficult task, for long. It was a matter of distances, and a matter of proper means of communication. It was only about a hundred years ago that this matter received considerable attention both in the UK and the USA. It was largely as a result of this investment of devoted efforts to address this problem that the agricultural advisory services in the UK and the agricultural extension service in the USA emerged as systematic extension education strategies.

Confining the attention, at this point, to the extension system, it is to be recognized that it is basically a system which functions as a link between the source of new knowledge and the points where such knowledge is put to use. This is the primary function of agricultural extension. Incidentally, it also functions, simultaneously, as a link that brings back the field problems for the attention of research agencies. This is the classical view of the agricultural extension system. It operates as a two-way channel, facilitating a flow of need-based, new knowledge from its source to the farming communities that have a use for that knowledge, as well as a channel facilitating a return flow of the farm problems, both prevailing at the time and those generated by the application of new knowledge, to the research centres for investigation. In this perspective, agricultural extension becomes a strategic, bridging mechanism between the knowledge centres and the farming communities using this knowledge resource.

Agriculture Extension is A Change Agency

In recapitulation, it is to be recognized that over the last two centuries, societies across the world are gradually differentiating into urban and rural segments. The urban communities are advancing economically more rapidly, relying upon trade and industry. They are becoming important knowledge centres, developing educational establishments, and learning from their contacts with the outside world. They are also on the forefront of cultural change.

In comparison, rural communities have remained backward. As such, the economic disparity between the urban and rural societies is widening. In recent times, the democratic regimes, in particular, have found it necessary to place a special emphasis on rapid rural development. Thus, being the major rural occupation, agricultural development receives the priority attention. Naturally, extension education becomes a central activity, in this approach.

What is Extension Education ?

There is a profound rationale behind extension education. The wealth of a nation is made up of two assets – the natural resources and the human resources. Nation's prosperity depends upon the development of human resources leading to the development of natural resources. One of the largest natural resources in the country is agricultural land. It is managed by thousands of farm families. Farming has been developed so far, using the traditional knowledge and practices. Further development in agriculture has to come from the application of modern science and technology. For this to happen, farmers have to be provided guidance, in adopting scientific practices in agriculture.

Generally, scientific knowledge in agriculture originates from the research centres. Therefore, there is a need for some agency that links the sources of new knowledge with the farm families. Agricultural extension is that agency, mainly. In this background, agricultural extension may be seen as a way of working with people to help them to bring about improvements in what they do. Such improvements become possible only when they adopt new ideas and practices suitable to their situations.

Extension is a change-inducing system, which enables people to improve their living, mainly through their own efforts. It leads to a betterment in their ways of living and making a living. By upgrading the knowledge and skills of people in what they do – their occupations – it helps them to increase their productivity, production and their incomes. In this sense, agricultural extension strives to help people to attain a more satisfying level of living. Its main route is informal education, in response to farmers' needs, problems and opportunities.

Extension education is defined in many ways. Here, is a simple, operational definition taken from Guide to Sericulture Extension : *"Extension education is a system of working with farmers, their families and communities, using informal educational procedures, to relate useful, practical knowledge to their needs and interests, thereby enabling them to make satisfying improvements in their living".*

This is based on a simple perception. People, as individuals, families and communities, always have some needs and problems. They are interested in meeting these needs and problems in order to create more satisfying situations. The extension agency helps them in this effort. It does so by introducing or developing appropriate new ideas and skills. Such innovations are made by way of informal education, and not by compulsion or coercion.

Components of an extension system

Every extension system is an outside agency. It works with local farming communities to promote improvements in farm production, family life and community living. It becomes eligible to undertake this task only because it has access to an accumulation of *new knowledge,* useful to farm people. This body of new knowledge is the starting point of all extension efforts.

Every extension system consists of five components. They include, (i) body of new knowledge, (ii) farming communities who need the knowledge, (iii) an organizational structure to link the two, (iv) a set of established communication methods, and (v) the infrastructure and support services. The five elements are briefly described below :

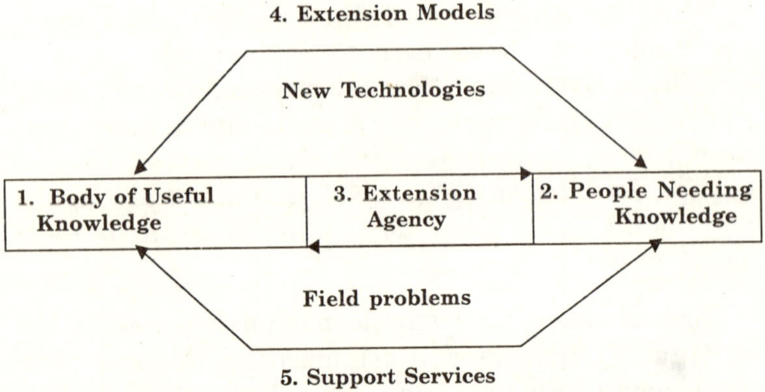

Fig 1. A Basic Model of Extension Education

(1) *Body of useful knowledge* : There is no need for an extension service if there is nothing to extend. As such, a body of new knowledge of practical usefulness is the first prerequisite in any extension work. Usually, generation of new knowledge is the task of the research system. In addition, in real life, useful ideas also come from resourceful farmers and the traditional sources. Such knowledge is gathered and validated to form the basis for recommended practices.

(2) *People needing new knowledge* : Talking of farming communities, one can recognize their common desire to improve their production and income. There is perhaps no farmer who does not want to improve his farming and his farm income. Farmers always have problems, needs and opportunities. Only, they do not often have an access to the new knowledge, or resources to secure and use such knowledge. Thus, if there are good ideas there will be people who can use them, sooner or later.

(3) *Extension agency* : Linking the source of useful knowledge with farmers who have a use for such

knowledge is the essential function of the extension agency. The task involved is not merely communication of information or transfer of technology, as is frequently assumed. It is delivering the *new knowledge* to prospective users and also helping them to assess and adopt that knowledge. The latter task is educational in nature. Besides, extension is a two-way channel, which not only brings new knowledge to farmers but also brings back the *field problems* for research investigation. Extension is, thus, a major vehicle for rural development.

(4) *Array of extension methods*: In developing educational interaction with farmer's, the extension system employs a combination of specialized and established communication channels called extension methods. There are a number of extension methods. They are generally categorized as individual contact methods, group contact methods and mass contact methods. Since the extension service works with farmers, many of whom are illiterate or semiliterate, extension methods make use of a variety of extension teaching aids. Field extension staff has to be trained not only in the principles and procedures of extension education but also in the selection and effective use of extension teaching methods.

(5) *Agricultural support services*: The primary endeavour of extension services is to bring about improvements in the existing production practices of farmers. This means bringing about changes in farming. Some of these changes merely involve the way things are done, while some others involve use of new inputs, which are often purchased materials and services.

The farm surpluses produced from this effort will have to be sold outside the farm. Thus, to sustain market farming, adequate infrastructure facilities must be created, including input supply agencies, marketing facilities, storage and processing arrangements along with the technical services.

How does extension system operate ?

The main task of extension education is to make available to people new perceptions, practices and products for which they have a need.

- It works as a system of informal education helping people to improve their living. It remains educational in its objectives, principles and methods.
- It develops informal working relationships with individuals and groups, studies their needs and problems, develops problem-solving approaches in partnership with them, and gets them adopted.
- It maintains links with the research agency in order to ensure continuous flow of useful information and to remit field problems for investigation.
- It fosters growth of the required infrastructure facilities to support development including supplies, services and marketing.
- It encourages farmers to evaluate the changes they have made to derive lessons for future application, and to derive a sense of accomplishment as a motivation for further effort.
- It encourages farmers to organize themselves into groups and associations to safe guard and advance their interests.

Good extension work benefits the community at large. Useful ideas and practices introduced by extension will gradually get established in the community. Directly and indirectly these improvements also get diffused in the region, becoming acceptable to more and more people.

It is also important to reckon what extension is not. Extension has its own scope as well as limitations. As such, it has to be used only for what it is suitable. Elsewhere it may be a misfit. We should recognize what it cannot do, specifically.

- Extension is not coercion. It cannot be used to push, pressurize or force people to do things against their will.
- Extension is not inducement. Sometimes, subsidies and incentives are linked with extension recommendations. This is a misleading approach. People should be able to accept a new technology entirely on its own merit and not because of the inducements.
- Extension is not salesmanship. For, a salesman works mainly for the benefit of himself or his firm; but extension works to help people as users of new technology, with their benefit in mind, all the time.
- Extension is not merely communication of messages. It goes beyond. It educates people not only about the usefulness of a new idea but also its application in the local situations. Thus, it does not merely inform them of the good ideas, it helps them to make use of those ideas.

Extension education is a twin process

There is another feature of extension education, which does not often get due attention. As an instrument of intervention it has two distinct aspects. One, it is a *process of communication.* Being a change agency, it has to be in continuous contact with client communities. As changes take place in the client system, needs may arise for new ideas and further changes. Thus, while the process of communication-based interaction continues, the messages in the communication, as a two-way system, will also keep changing. This is the communication aspect of extension education.

Second, it is a process of *adult education.* As a change agency, it interacts with adult farmers. These adults would have reached a level of maturity, and would have equipped themselves with the knowledge and skills required for meeting most situations in daily life. Often, they would have gained a conviction that "they know all that they need to know in life." Therefore, adults to be good learners, have to unlearn many things. In dealing with this situation, extension education must have adequate competence in the appropriate methodologies pertaining to adult education. From this perspective, considering extension education merely as a process of communication will be quite erroneous.

Extension intervention is, primarily, to produce changes in farming practices. These changes, in turn, produce other changes, some anticipated and others unanticipated. As a next step, extension has to deal with these two kinds of changes as well, as necessary. Then, when one set of needs and problems of farmers are met, there will normally be other needs and problems waiting to be given the attention. Also, as extension serves one generation of farmers, there will be another generation of farmers who have to be attended. Moreover, new knowledge itself is considered to be continuous in its emergence. For all these reasons, extension education is regarded as a service required to be available to the farming communities continuously.

The process of change in the farm sector will normally be slow, at least in the early stages. Theories concerning innovation, adoption and diffusion in a social system have been widely validated. The pace at which new ideas get accepted and shared will, however, depend upon a cluster of socio-cultural factors, stage of development as well as the characteristics of the innovation.

Generally, extension activities are conducted with four prominent outcomes in view. One, the intervention should make a positive impact, through new knowledge, on the ***production efficiency*** in farming. Two, it should result in ***economic improvement*** in the farm sector. Three, it should gradually lead to ***capacity building of farmers,*** as individuals, groups and communities, as a measure of human resource development. Finally, it should culminate in the ***institutionalization of development,*** as a manifestation of the farming communities internalizing the new ideas and practices.

The Changing Agricultural Scenario

In the last fifty years, from 1950 to 2000, tremendous changes have taken place in the agriculture sector. Very conspicuously, cereal production has gone up from about 50 Mt to 210 Mt. Irrigation has expanded considerably. Market growth is seen in segments like milk production, horticulture, sericulture and floriculture. At the same time, some disturbing trends are seen in declining yields both in dry and irrigated farming, depletion of the vegetative cover on and around the farms, and sinking ground water levels. Both these positive and negative changes have implications to agricultural extension work. With this viewpoint, a brief account is given below of the more prominent changes in the agricultural sector, which have a direct bearing on the extension tasks of the future.

Green Revolution and its impact

A brief look at the gains and losses accruing from the Green Revolution, in broad terms, will serve here as a useful backdrop. With Freedom, we inherited a serious food deficit, along with a stagnant agriculture, embedded in traditional practices. It was a mixed farming system with livestock and tree crops as components. It was entirely organic farming, pursued as a family occupation with subsistence as its goal.

But, in meeting a threatening food scarcity, Indian agriculture had to move rapidly. Vigorous efforts in the form of the Grow More Food Campaign were mounted in the wake of the War years, with emphasis on expanding irrigation and mobilizing local manurial resources including oil cakes. Its impact, however, was limited. But, by the mid 1960s, a new era dawned on the agricultural scene. New biological material in the form of hybrids and high-yielding varieties in major food crops entered the farm front. Simultaneously, high-grade, complex fertilizers also became available. Availing these powerful inputs, progressive farmers, in favourable farming situations, began making history by doubling the crop yields. This innovation has come to be known generally as the Green Revolution, based on the "seed-fertilizer" technology, mainly impacting crops and yields in traditional farming. Within about two decades, the country not only achieved food sufficiency, but was also able to produce modest surpluses for buffer stocks and token exports.

When we look back, what were the major gains? Primarily, the mental horizon of the Indian farmer got considerably widened. Millions of "illiterate Indian farmers", steeped in tradition, living in isolated communities, were found to accept the most advanced farming technologies of the day, and adopt them with a fair degree of efficiency. The farm sector benefited from the sophisticated crop varieties, fertilizers and pesticides, as well as advanced agronomic techniques and managerial practices. Thus, farmer, farming and farm-support services began moving beyond the traditional system.

But, what were the losses? Of course, there were some losses that were inevitable, and some more that were avoidable. The localized, self-reliant family farming system yielded place to the surplus-producing, externalized farming system. In setting his farming goals and making decisions on the production practices, the farmer lost a good deal of his independence. Also, at the same time, some of the sound traditional practices like mixed farming, mixed cropping and crop rotations gradually yielded ground to systems of mono-cropping. Thus, once for all, farm families became dependent on the larger economy and outside world, for making a living.

Then, what are the things overlooked? In making a hurried progress, the Green Revolution lost sight of many critical aspects of a healthy farming system. As the end-users of the natural farm resources as well as modern technologies, farmers were rarely guided, in an adequate measure, beyond crops and yields. Thus, due attention was not given to matters like conserving, developing and judicious utilization of the natural farm resources, maintaining adequate environmental support to farming, planning production of surpluses and dealing with externalities including management of market operations. Also, as subsistence farming moved into surplus farming, then into market farming, institutionalization of several phases of agricultural development became critical, but was hardly given the needed attention by the Green Revolution.

When all this is put together, it becomes obvious that the Green Revolution was essentially a yield-enhancing modern production technology, and nothing more. On the other hand, a more comprehensive agricultural development approach should have devoted attention, simultaneously, to natural resource management on the farms, and to build a more adequate environmental support to farming. Also, attention was required to strengthen the backward linkages including dynamic research and extension agencies and dependable input supply systems, along with the forward linkages including economic incentives, market infrastructure, agro-industries and domestic and export trade facilities.

Purpose in farming is changing

The farming goals are changing. In the country as a whole, the farming system is changing in many ways. This is basically for the reason that the very purpose of farming is in transition, over the years. Farming is no more a pursuit for subsistence. There has been a vast transformation, arising from both the development initiatives, and the autonomous trends. This change is taking place in discernible phases.

The occupation in the past was mainly ***family farming*** for subsistence purposes. It was entirely a self-contained system, with farming knowledge coming from the older generation, seed from the previous harvest, manure from the farm wastes, and labour from the joint family, while the produce was meant primarily for family consumption. The occasional surpluses, caused by favourable seasonal conditions, were saved for a lean year. Then, faced with an enormous food deficit, in the early years of Freedom, the country had to adopt ***surplus farming,*** making use of the Green Revolution technologies. Here, apart from some new technologies, the inputs like seed and fertilizers also came from outside sources.

During this period, rural society was also moving away from barter economy into money economy, as the system of village services and rural artisans began decaying. As such, all farmers, big and small, started taking to some kind of **market farming**, seeking cash incomes to meet family expenses on one hand, and to purchase farm inputs on the other. There were broadly three kinds of market farming ventures. One, a large number of small farmers, occasionally using purchased inputs, produce commodities mainly for the local *shandies (village / block level local market).* Two, a small number of elite farmers take to high-tech specialized farming to produce exportable commodities of international standards, against known demands. Three, a considerable number of mid-size farmers with a good access to technology, resources and political power go in for producing huge surpluses of common commodities, unmindful of the limitation of market demand. It is mainly they who create the well-known market gluts and price slumps.

Then, the recent trends of liberalization, privatization and globalization are now pushing farming into international market competition. Thus, **commercial farming** has become inescapable. Here, farming has to be conducted as a business. The saleable farm surpluses will have to face global competition not only in the international market, but even in the domestic market. In this context, the farm output for the market must be competitive "both in quality and in unit price". Particularly in doing this, we suffer from two great handicaps — we have a small-farmer agriculture, which denies us the benefit of **economies of scale,** and the bulk of our farmers have yet to acquire the **managerial abilities** that are required in conducting farming as a commercial business.

As of now, Indian farming is still a dual-goal pursuit, serving the purpose of both food security and income security, for the farm families. All the same, Indian farming is increasingly becoming dependent on external factors like technical information, production inputs, price support and market facilities. As a result, the success in farming now depends, not only on the farmer himself but also on many externalities, much of which is beyond his comprehension, let alone his control. Hence, the new development approach must help farmers to understand the nature of the larger economic world, of which our farming is becoming an integrated part, and to learn to survive in the new environment.

Subsistence (Main features)	**Commercial (Main features)**
Food Security	Food, Income
Mainly Traditional	Mainly Research
Local Inputs	Purchased Inputs
Very Little	Substantial
Moderate	High
Local	External
Low	High
Little	Very high

Fig 2 Changing Features of Farming

Degeneration of Farm Production Base

Land-man ratio narrowing

India happens to be one of the most populous countries in the world. In the last fifty years, its population has grown more than three-fold. The rural population alone has increased from 280 M to 742 M. In the meanwhile, for the country, the total land area and the cultivated area have almost remained the same, at 329 M ha and 142 M ha, respectively, with only around 30% of the sown area having irrigation. This situation has put a tremendous pressure on natural resources in terms of land-man ratio. As such, of nearly 116 M farm holdings in 1995-96, around 71 M are found to be marginal, 22 M are small and 21 M are medium and only 2 M are large. The farmland per capita, which was 0.46 ha in 1951, became 0.26 ha in 1981, and may have come down to 0.15 ha in 2000.

Vegetative cover shrinking

India, as a subcontinent, has 15 agro-climatic zones. Depletion of key natural resources like soil, water and biodiversity is taking place in all these zones. The inadequate forest cover, against a norm of 33%, is a matter of concern. According to the State of Forest Report 1999, total forest cover was only 19.39%. Out of this, dense forest (40% crown density) was 11.48% and open forest (10-40% crown density) was 7.76%, while mangroves was 0.15%. Even the rest of the non-arable land is largely depleted of its vegetation. What is involved here is not merely the lack of sufficient plant cover, but, more importantly, loss of biodiversity, in addition to the impaired microclimate and bio balances, and loss of soil, rain water and organic matter.

Land productivity declining

On account of extensive loss of vegetative cover and poor conservation efforts, farmland has suffered severe erosion. According to one estimate, annually, about 5300 Mt of topsoil is lost through erosion, which is about 12 t/ha. In the process, the three major plant nutrients — nitrogen, phosphorus and potassium — ranging from 5.4 to 8.4 Mt are also lost, every year. Further, there is one to two percent loss of storage capacity in the tanks and reservoirs, per annum, due to silting.

There is another aspect of productivity loss in farmlands. Degradation of land in the country is put at 187 M ha. It is estimated that soil degradation due to water erosion is about 148 M ha (57%), due to wind erosion is about 13.5 M ha (4.1%), chemical degradation is about 13.8 M ha (4.2%) and due to water logging is about 11.6 M ha (3.5%). The cumulative effect of this degradation of natural resources further leads to an enhanced exploitation of the declining resources by an expanding population. It may not be possible to stop these deteriorations completely, but surely they could be minimized.

Water resources dwindling

Assuming that the amount of annual rainfall received, on an average, remains the same, the competing demand for this limited resource is constantly increasing on account of a growing population, changing life styles, expanding irrigation and enlarging industrial and urban sectors. The water-use efficiency, particularly under the public irrigation systems, is considered to be very low, at 30%. The over-use of ground water in many regions, beyond the limits of recharge, is posing serious environmental problems.

Even then the management of rainwater leaves much to be desired. R.S. Paroda (1998), for instance, opines that, while the country has a fairly adequate rainfall, water harvest and utilization patterns have not been sufficiently evolved. The picture is this, in terms of million hectare metres (Mha m) :

Annual rainfall 400 M ha m
Rainwater going into rivers 186 M ha m
Rainfall on farm lands 160 M ha m
Rainwater before and after crop season 40 M ha m
Rainwater available for harvesting 24 M ha m

The combined effect of these various aspects of the farm sector is something that should matter in formulating a comprehensive agricultural development approach. While it has to be readily conceded that all these different factors can hardly be tackled together or simultaneously, one has to keep in view their inter-relationships, so that combining or sequencing of the development efforts would have an add-on effect over a period of time.

With the unabated population growth, fast-changing life styles and easy access to powerful technological tools, the manner of resource utilization in farming is rapidly changing. Extensive exploitation of ground water is a case in point. In the last one generation, in many areas, groundwater levels have fallen from about 10 m to 200 m. Farmers ending up with dry wells is becoming far too common. This has several consequences. Dead investments are on the increase. Viability of commercial farming is threatened. Availability of drinking water becomes a serious problem. Tree vegetation comes to suffer, since the root zone of about 10 m is getting drier every year. Tragically, this is a man-made crisis.

Farmers feel trapped in farming

Rural aspirations are rising

In the last fifty years, the isolation of village communities is getting reduced, with improved road links and transport facilities, increased literacy levels and expanded media reach. Along with this kind of greater exposure to the outside world, the expectations of rural people have also risen, particularly with the rural youth. Many have even begun to feel that they have been denied a fair share of the public good. Thus, the effect of a "shrinking world" is becoming conspicuous. The mindset of the farmers too is changing, regarding the farm work as drudgery, and farming itself as a non-paying proposition. Farm life seems to have lost much of its former glory.

Development support weakening

Technology is often seen as the engine of growth. Indeed, it was the new agricultural technology that powered the Green Revolution. But, by the 1980s, the outflow of new technology of a substantive nature, in major crops, from the public research centers, began dwindling. Even though some new technologies did emerge in the case of subsidiary enterprises like poultry, dairy, sericulture and vegetables, they were hardly of benefit to the majority of farmers. This is often attributed to slackness in the research and extension systems, mainly due to lack of policy focus and development thrust.

Of course, "leaps and lags" are common in research outflow. But, in the wake of the Green Revolution, the pace of development need not have slackened so much. There was plenty to be done as follow-up work, after the Green Revolution. The full potential of the new technology had not been tapped in many cases. Adaptation of the new technology to the less-endowed farming regions still remained to be attended. Also, while the Green Revolution technology had only the "crop-yield" focus, diversification of farm enterprises offered an enormous scope for further development initiatives.

Further, investments in the farm sector, both public and private, markedly declined in the later years of the 20th century. Beyond achieving food security, as seen earlier, there were a good many things to be done in the agriculture sector, with faming moving towards market orientation. The status of soil, water and vegetative resources could have been in a better shape today had timely attention been paid to their conservation and development. Similarly, building up more adequate and dependable backward and forward linkages should have received a greater attention.

Rural economy lagging behind

In the early stages of development planning, it was the anticipation that, with a rapidly expanding economy, some of the surplus rural population would get absorbed into the industrial, trade and service sectors, thus improving the land-man ratio in the farm sector. But, this did not happen, due to the undiminished population growth, on one hand, and on the other, the industrial sector becoming more technology-intensive rather than labour-intensive. With shrinking farm holdings and rising production costs, farming became a less remunerative pursuit. Rural employment also did not expand very much. Thus, rural incomes remained far behind the urban incomes.

Meeting the WTO challenges

Trade between countries is the focus of WTO. It is not so much the production matters. Relative to the trade in farm commodities, there are three basic elements in the Agreement on Agriculture (AoA) that govern the transactions — production subsidies, export incentives and import tariffs. Recently, India removed the quantitative restrictions (QRs) altogether. Thus, today we face a free trade situation between countries, of course, as manipulated through the domestic support and import levies.

It is seen that the developed countries presently enjoy an unmatched advantage under this arrangement. In simple terms, having expanding economies relative to shrinking farm sectors, with dwindling farm populations and declining farm share in GDP, they are in a position to provide heavy domestic subsidies. But, this is not the case with the developing countries, like India. As a result, developing countries are faced with a situation where highly subsidized foreign farm goods enter their domestic markets, which they seek to counter mainly with import levies, within mandatory limits. One graphic example reported in the media, soon after Cancun meeting, was that a specific farm commodity costing $10 in the developed countries, compared to $7 in the developing countries, was supported with a $5 subsidy to make it competitive, internationally.

There is another, more general point. Farm commodities compete in the market, on the basis of their quality and cost. These attributes are not imparted to farm goods, after they arrive in the trade sector, but in the farm sector, where they are produced. Therefore, for Indian farm goods to be competitive, appropriate attention has to be devoted to production management in the farm sector itself. So far, this was not a matter of concern to the Indian farmer. However, to ensure competitive quality and cost in the farm goods, farmers have now to gain the necessary knowledge and the required managerial competence.

But, the average Indian farmer is hardly in such a position. Let us look at the cost factor. It is here that the concept of "level playing field" needs to be extended beyond the trade sector, into the production sector. Everybody knows that Indian agriculture is made up of small farmers, with about 78% having less than 2 ha of farmland. Then, how can these small farmers compete in the market place with the Western farmers having 200 ha of farmland? How can these farmers avail of the **economies of scale?** Thus, in terms of attaining competitive unit costs, there is surely no level playing field as of now.

Let us also look at the quality factor. The Western farmer, in the post-war years, has steadily moved towards specialized farming as a commercial venture. He now manages farming almost as a business undertaking. As such, he has become far more competent as a farm manager. He enjoys a highly dependable input supply system, technical service and market infrastructure. Moreover, he has become a part of well-established network of farmers' associations that strive to protect their common interests. In comparison, the Indian farmer, most often, stands alone, deprived of this kind of support. As a result, as a manager of farm business, he remains essentially a traditional farmer, just moving into market farming, and is nowhere near a competent, Western commercial farm manager. Hence, the quality factor is not yet within his reach.

In this background, it is obvious that the average Indian farmer is handicapped in meeting the international competition, in the absence of strenuous efforts for adequately equipping him for this purpose. Hence, the future agriculture development approach should necessarily focus on preparing the farmer as a manager of commercial farming, build adequate infrastructure needed for a commercial farming system and promote producers' associations not only to protect their interests but also to overcome some of the limitations of small holdings, through strategic cooperative efforts. This is, indeed, a tall task. But, there does not seem to be any escape from this hard, complex alternative.

Survival requires efficiency

An overview at this point will be useful. Farming is all about generating certain goods we need, through selected biological processes, harnessing the available natural resources and applying the production technologies evolved for the purpose, either traditionally or scientifically. In recent years, however, we seem to have devoted more attention to production inputs and technologies, and not so much to management of natural resources. This has begun to hurt our farming capability, in the recent years.

When it comes to natural-resource management, the primary users of soil, water and biodiversity are farmers and farming communities. The conservation and development of the on-farm resources is the direct responsibility of farmers, whereas the responsibility for off-farm resources, reserve forests apart, is that of the farming communities and village institutions. At present, there are marked deficiencies in managing the on-farms, and more so of the off-farm resources.

In the case of on-farm situations, efficiency has to be increased on two fronts. One, the care and management of natural farm resources in terms of soil management, water use efficiency and biomass production, as of today, leaves much to be desired. Two, there is considerable room to improve the productivity in many chosen farm enterprises, by way of upgrading resource-use efficiency, in addition to fuller utilization of the potential of the new technologies. Hence, the attention, now being devoted mainly to crops and yields, must be intensified and extended to farm-resource management also.

Adequate environmental support to farming in semi arid regions is very critical. Off-farm vegetative cover helps in better management of rainwater, as it softens the impact of rainfall and impedes surface flows, thereby reducing erosion and increasing infiltration. It improves microclimate, and enhances atmospheric moisture. It serves as a source of biomass for use as fodder and manure. In addition, it improves the biological balance between crop land and the surrounding areas, with increased bird, bee and insect populations. But, farm surroundings today stand depleted of their vegetative cover.

Since farming is no more a simple family pursuit, and has become a prominent production sector in the national economy, there are other phases in the development process that require serious attention. Present-day farming has become dependent on many external factors. It has to be supported by certain backward linkages like supply of production inputs and new technologies. Likewise, it needs forward linkages in the form of roads, transport, store-houses, processing and market facilities. A stage in agriculture development has now been reached where the on-farm efficiency critically hinges on the adequacy, dependability and cost-effectiveness of these linkages. Further, in the long-term interest of agricultural growth, a conducive economic climate has to be created by way of credit support, remunerative prices, insurance and strategic subsidies.

Farmers' associations and village organizations are of critical importance to development. Since top-down approaches have failed in the past, promoting farmers' initiatives in development may prove better, as in the case of self-help groups. This effort is very necessary for enabling farming communities to become sensitive to market forces, capable of planning commodity production and organized to deal with hostile, outside influences. Thus, the ability to work together as producers' associations to make collective decisions and act jointly will become a strategic necessity in the future.

Farmers taken for granted in development

Agriculture development is not an undifferentiated, 'single phase' process. A closer look will reveal that it is made up of at least two distinct, but inter-related phases — the *production phase* on the farms, and the *enabling phase* beyond the farms. The policy makers and programme implementers operate most of the time on the impression that creating certain enabling conditions, through specific externalities, commonly recognised as backward and forward linkages, is an adequate impetus for agriculture development to take place. The expectation is that these incentives and opportunities will function as the "pull" and "push" factors, motivating the farming communities to enhance their farm production as well as their production efficiency.

This assumption is correct, but only partially. The farming communities are not homogenous social systems. They are made up of, in classical terms, innovators, early adopters, early majority, late majority and non-adaptors. As such, those who really make use of these incentives and opportunities are the innovators, and resourceful farmers. They happen to be a small fraction of the farming community. But, the bulk of the farmers are not like them. They are mainly resource-poor, traditional farmers.

At this point it is necessary to recognise one more feature of the agriculture sector. It has a two-fold function. One is to ***generate wealth*** in this primary economic sector. This is the role performed by the elite farmers. Second is to ***provide livelihoods*** to the vast rural population. This is a core objective of the development undertakings. Hence, in the agriculture development approach, the livelihood goal is as important as the wealth creation goal.

In considering the on-farm production phase in agricultural development, it is essential that certain significant variables are taken into consideration. Farming, after all, is what the farmer does. What the farmer does will be based on what he knows, what he believes and what he manages. When we take this view about farming, we come to the realization that the ultimate decision-maker in farming is the farmer himself. All the rest, like the policy makers, development agencies and service providers are just the enablers and supporters.

The elite farmers, as keen information seekers, will be able to quickly take advantage of the development incentives and opportunities, mostly on their own. But, the non-elite farmers, whose welfare is one of the core objectives in the development policy, have to be ensured of the farm-door extension and support services. In the case of both these categories of farmers, there exists a huge **capacity building task,** as a basic responsibility in agricultural development. But, this significant matter is yet to get adequate recognition in policy formulation.

As indicated earlier, farming is an enterprise that requires **management.** It is easy to accept that the kind of managerial ability required for commercial farming is not the same as what is required for subsistence farming. This means that, as the farm enterprise undergoes a transformation, the **managerial competence** must also undergo an upgradation. Further, it is not difficult to accept that while the elite farmers are quick to find ways to improve their managerial skills, other less endowed farmers, may not be able to do likewise. The implication is clear – agriculture development strategies must have a strong component of **capacity building** in the case of both the middle and bottom segments of the farming communities.

It is evident from the foregoing that the context in which extension system is required to operate has undergone an enormous change, both in breadth and depth. In this regard, as relevant to the extension intervention, what are some of the specific circumstances that need to be taken into reckoning ? In exploring this matter, the focus here will be mostly on dryland farming situations, since both in the land area involved and the dependant population, it happens to be the major segment in the agricultural sector.

One of the basic tasks of agricultural extension, as seen already, is to help people to improve the production efficiency in farming. We see that the simple "seed-fertilizer" days in farm development are long over. Agricultural development has become a vastly complex undertaking. Today, two serious on-farm problems have emerged to demand our attention. One on-farm problem is the extensive ***mono cropping practice,*** with its consequences, has become a major concern. In situations where holdings are becoming smaller, farmers resorting to grow, year after year, the same commercial crops like groundnut and cotton. This has led to rapid soil exhaustion, resulting in pronounced micro nutrient deficiencies. Build up of serious pest and disease problems have also become common. Since continuous mono cropping is leading to declining yields, farmers have resorted to application of higher doses of some fertilizers, but not in a balanced combination.

Another on-farm problem is related to the inadequate attention given to ***natural resource management*** on the farms. For instance, *in-situ* rainwater management, as it is practiced today, is not as good as it was in the olden days. As such, even when the total annual rainfall is large enough, relative to the requirement of a monsoon crop, management of soil moisture during the dry spells is not as good as it could be. Similarly, maintenance of soil fertility levels, particularly by using adequate amounts of organic manures, leaves much to be desired. A solution to this problem is generating alternate farming practices.

Then, there is a critical off-farm problem. It is the serious depletion of the ***environmental support to farming.*** Particularly in dryland farming situations, with most of the sown area in the country falling under semi arid conditions, the presence of vegetative cover around croplands is very important. This vegetative cover contributes to higher atmospheric humidity levels, reducing the surface evaporation in cropped areas. This also prevents gusty wind flows, again lowering the evaporation losses in cultivated lands. Further, the plant cover on the non- arable lands serves as a source of biomass for fodder as well as for organic manures. More over, such vegetated environment will provide for better bio balances by way of increased bird, bee and other insect populations. In addition to these, as a measure in managing the rainwater on the non-arable lands, which amount to nearly half the geographic area, the vegetative cover serves to impede erosion and enhance infiltration. In recent years, this relationship between the cropland and its environment has been largely overlooked. Thus, there is scope for building the awareness of rural communities in this regard.

Apart from addressing these on-farm and off-farm distortions, the second important extension goal is ***facilitating economic improvements*** in farming. Today, farming requires money investments, coming out of either the savings or from the borrowings. Ensuring cash income from farming, therefore, has become a necessary part of the pursuit. However, extension system has not devoted the required attention so far to treat farming as a business undertaking. As such, it does not even consider bringing into focus the economic aspects of the farm improvements it recommends to the farmers. Very evidently, it does not help farmers in planning their credit borrowings, in using the credit properly or in repaying the credit. As such, one of the most necessary educational services required by farmers, in investment–oriented farming, the extension system is not able to provide.

The third basic extension responsibility is ***capacity building of farmers,*** as individuals, groups and communities. As farmers enter into commercial farming, they have to take into account many more factors in their enterprises As an example, market farming requires that farmers have the basic knowledge of the market behaviour and the demand patterns for farm products. On a broader plane, they should be able to take into reckoning the technological, economic and political changes occurring in the world around them. They should also become capable of identifying and interpreting the implications of these changes, and make appropriate adjustments in their production ventures.

Lastly, the fourth extension function is ***institutionalizing development***. Since all development is change-based, and most changes have to be durable, it is necessary that the continuity of the changed features is adequately based in new social structures and processes. Thus, the external linkages in commercial farming, both for dependable supplies and services and for market support, have to be properly established. Also, since Indian farming is largely made up of small farmers, groups and associations, mainly as regional or commodity bodies, have to be fostered. Such associations have become necessary to safeguard farmers' interests on one side, and to enhance their bargaining capacity in the market place, on the other. This constitutes the human and institutional development aspects in modernizing agriculture.

In retrospect, it is seen that even as the farming occupation changes, the basic responsibilities of agriculture extension will continue to be the same. Of course, the kind of activities pursued under each of these responsibilities will change. Thus, while the role in improving the production efficiency in farming remains in tact, the technologies dealt with will keep changing. While the role in improving the economic efficiency in farming remains in tact, the issues focused upon will change. While the role in capacity building remains in tact, the focal tasks will be changing.

Similarly, while the role in fostering institutionalization of development remains the same, the kind of social structures that become critical will be changing. Further, as new generations of farmers take position in farming, and as the traditional farm knowledge requires to be supplemented with external sources of information, extension education will continue to perform its main roles, while the content and focus of these role will change from time to time. .

Prospective Extension Tasks

As seen so far, the farm sector has undergone vast changes in many dimensions. As a consequence, while the legitimate responsibilities extension system remain the same, the set of tasks it has to perform in each of these dimensions will have to be substantially different, consistent with the changing needs of the time. How does agriculture extension go about finding ways to remain relevant in this dynamic process? Here is a visualization.

Farmers are a changed clientele

The extension system, at present, has to deal with the third generation of farmers since Independence. Today, they are more educated, more mobile, and also have a wider horizon and broader range of contacts. They have different life styles and higher aspiration levels. In a society that has gone through five decades of democracy, they seem to have gained a new sense of their rights and privileges. Thus, they are now an entirely a different kind of extension clientele, compared to farmers two generations ago. The extension system, therefore, has to adopt a different approach of interaction with them.

In the earlier decades, during the Green Revolution period, there was a serious food deficit problem. Everybody was concerned about it, including the farmers. In this background, the new technology was seen as a great opportunity in the context of the traditional farming system. Hence, at that time, somewhat of a "prescriptive mode" of extension approach was found acceptable to the farmers. But, as the nation achieved food security, the flow of new technologies also began to dwindle. Meanwhile, the money economy having entered the rural life, farmers started moving towards market farming and diversification of farm enterprises. The extension system began losing its influence, somewhat. At this stage, farmers, on their own, began looking for new ventures like dairy farming, poultry production, vegetable cultivation and sericulture. The extension system, wherever it continued working with farmers, had to adopt a "participative mode" of extension work, where ideas and strategies were finalized based on shared thinking. Gradually, some elite farmers moved into high-tech farming and gave room to a "consultative mode" (where ideas and strategic means were decided based on expert opinion) of extension approach.

Viewed from another angle, this transformation in the role of farmers in their interaction with the extension system is in fact a natural corollary. Indian village communities in the pre-colonial days were widely known as self-governing social systems. That was the traditional socio-economic era. But, today, under a globalized economic order, with the local farming enterprises tied up with the larger external economy, modernization of farming has to begin mainly with a purposeful external intervention. Agriculture extension service is one such intervention. However, over a period of time, the farmers are likely to become increasingly capable of managing by themselves many more things in their life. According to this perception, the relationship between the intervention agencies and the local communities will pass through a development process that is seen as ***prescriptive to participative and finally to consultative.*** Extension system, as a friend, philosopher and guide, has to enable the farming communities in this process capacity building.

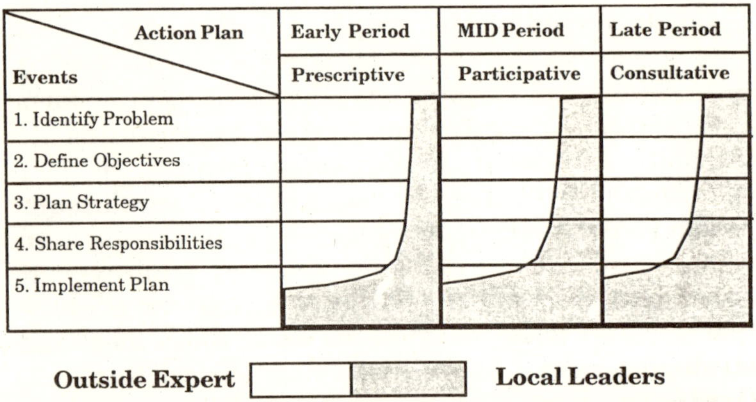

Fig.: Changing Roles of Outside Experts and Local Leaders

Extension support to dual-goal farming

The agriculture sector, as the primary economic sector, is serving two fundamental national purposes — capital formation and providing livelihoods. Over the years, for the reasons seen earlier, the farm sector has gradually differentiated itself into two distinct segments – one is a rapidly emerging ***commercial segment,*** and the other is a slow-moving ***subsistence segment.***

The commercial segment is made up of relatively small proportion of the farm population. They are venture-some farmers, constantly looking out for new ideas, with an eye on the profits. This is an enterprising lot. They also consider it as a matter of prestige to be on the forefront of the agricultural modernization process. Being resourceful farmers, with an innovative mentality, they focus on high-tech, high-investment ventures, taking reasonable risks. They are in fact a superior kind of managers among farmers.

The subsistence segment, on the other hand, is made up of a large proportion of the farm population, mostly of small, marginal dryland farmers. Ensuring the food security for the family engages a major part of their time and energy. Also, since the present day farm families must ensure a dependable money income also to meet their family and production needs, they are compelled to combine a small component of market farming along with food farming.

These are, again, resource-poor and risk-shy farmers. They tend to adopt only those new practices that are proven in the locality. In enabling them to adopt new practices on their own, they need a lot of persuasion. Since they form the bulk of the farm population, and since it is their livelihoods that are at stake, the extension system is required to devote a considerable amount of attention to this segment. Also, as farm managers in a changing environment, they are often found to be average. Hence, ways have to be found by the extension system to become more effective in working with them than in the past.

In formulating the extension strategy, a key factor is that farmers as adults differ in their background, understanding and learning abilities. From this viewpoint, farmers in the commercial and subsistence segments have to be treated as different, in many ways. Another key consideration that requires attention is that the technologies vary vastly in their nature. While some are simple, others are complex. As such, it is necessary that the extension approaches are substantially different for the commercial and the subsistence segments.

Specialized Extension Service

Supporting the commercial segment, which entails high value enterprises, superior technologies, large investments, better management and rigid production standards requires an extension agency that is able to offer high grade technologies, prescribed standards, dynamic approach and dependable service. The present day public extension system does not measure up to these attributes. The chances are that such quality services are afforded mainly by the private extension agencies, as consultants who survive on their performance. Since most of the commercial ventures are specialized operations, this extension service must also be specialized agency. The task involved is one of providing guidance on venture-specific technologies and related managerial issues. Often, these clients may require guidance on input sources and market options as well.

The beginnings of such specialized extension services are already in evidence in ventures like poultry production, dairy farming, sericulture and floriculture. Professional extension groups, input agencies, farmers' organizations, trading houses and agro-industries are already into this business. Most often, these services are paid for.

It is critically important that this kind of extension service maintains a wide professional contact not only with the local and regional organizations but also with related national and international bodies, to stay abreast in the field. Then, it is also important that the national agricultural policy provides scope for promoting and fostering this kind of specialized extension service, as a development input. The premier agricultural agencies like the Ministry of Agriculture, ICAR and SAUs, must ensure both the professional manpower and the technical back up.

General Extension Service

For a long time to come, till the population growth rate relaxes, and till the trade, industry and service sectors of the economy are able to absorb the surplus rural population, the country is bound have a large subsistence segment in agriculture. Farmers in this segment will depend upon farming both for food security and for modest money incomes. Since these farmers are stuck with farming, with nowhere else to go, farm improvements, with the guidance of the general extension service, among others, will remain the main means to their livelihood improvements.

The task involved here is to enable the large mass of subsistence farmers, in a variety of farming situations to progressively improve the efficiency in their food crops as well as in the chosen market crops. Especially for the reason that they are low-income farmers, they will not be in a position to pay for the technical services they need. Also, since most of these farmers are engaged in earning a living every day, they may not be able to leave their farms and homes during the agricultural season, seeking new technical knowledge in distant places. It is for this reason that for these farmers, there should be farm-door delivery of the extension and support services. In a democratic system, this is a way of ensuring the social welfare services.

Focus on natural resource management

A point that should not be missed by extension education is that in the history of Indian agriculture, though it occurred as a rare and fortunate event, the Green Revolution has left behind in its wake many maladjustments in the farm sector. These distortions have to be properly addressed in the future strategies of agriculture development. The extension system has a key educational role to perform in this regard.

Mending the prevailing distortions in resource-use pattern is a major task. The soil, water and bio diversity under a climatic situation are the non-replaceable natural resources. In properly managing these resources there are two phases – natural resource management **(NRM)**, and natural resource utilization **(NRU)**. Many of the present day resource-use patterns, if they remain un-amended, may ultimately lead to the "desertification." Mono cropping in dryland farming, over-irrigation under canal systems and over-exploitation of ground water are a few cases of note. Building the awareness and capacities of farming communities in dealing with these problems is an urgent need.

Dryland farming has come to suffer due to mono cropping in cash crops, among others. The compelling factors of shrinking farm holdings and the need for money incomes, seem to have converged here. This land-use pattern has led to conspicuous problems like impaired nutrient levels in the soil and build up of un-manageable disease and pest complexes. In meeting these problems, a necessary measure is to adopt alternative cropping patterns involving the proper choice of the crops grown, crop combinations and sequences, and crop production technologies employed.

With nearly half the total geographic area opened up for cultivation, coupled with extensive depletion of on-farm and off-farm vegetation, the crop lands have come to suffer unabated water and wind erosion. It is reported that the extensive bare lands are losing large amounts of finer soil fractions annually. This will result in croplands becoming more sandy and shallow, every year, affecting crop production. This is largely avoidable if the farmers are enabled to restore some of the past land management practices along with the modern methods.

Similarly, the vegetative cover and the surface water bodies on the non-arable lands have to be built back as environmental support to farming. This is a task that involves action on the part of the farmer groups and communities. The Panchayat Raj (local self governments at village, block and district levels) institutions, community based organizations and NGOs have an important role to play here. Towards this end, the extension system has a social education role to perform.

Market operations, a critical capacity

Every farmer today has a need for money income. He will, therefore, plan to produce some saleable farm surpluses. It may be for the local market, or for the regional market. In cases where farmers go in for permanent cash crops like areca nut, coconut, coffee or vanilla, in a kind of specialized venture, they almost entirely depend on the market behaviour for the sale of their commodities. If there is a spell of low prices, their only option is to cut the production costs wherever they can, apart from resorting to some intercrops. However, most farmers go in for short duration seasonal cash crops like ground nut, cotton or vegetables. For these farmers too market slumps and price crashes are quite common, particularly in the case of perishables.

While the plantation farmers already know quite a good deal about the market behaviour and about the pattern of movement of their commodities, the bulk of the small producers of seasonal commodities know next to nothing about their markets. The producers of the perishables happen to be the worst lot.

Under these circumstances, knowledge about market operations becomes as important as the knowledge about the new technologies and purchased inputs. They have to become knowledgeable about the market networks, seasonal price patterns and demand manipulations by middlemen. Born out of such knowledge, farmers should become capable of making appropriate decisions regarding their production plans. Therefore, the extension efforts are necessary in the future to educate the farmers about the market system, demand-supply relationships, price fluctuations and related matters.

Related to the market issue, there is another important matter that requires serious extension attention. For the farm products to be competitive in the market place the goods must measure up in two significant features – one is the ***product quality*** and the other is the ***unit cost.*** The quality issue, the farmers can handle as their ***managerial competence*** improves. As for the cost issue, ours being a small farmer economy, it cannot be handled through the managerial improvements only. For, the factor is directly related to the ***scale economy.*** The option available in this case is adoption of joint efforts by farmers' groups wherever possible. Group farming approach, in a way, may be inevitable for us in coping with the globalization effects. Commodity associations or regional associations for chosen commodities may be helpful not only in production planning and safeguarding common interests but also in cutting costs in input purchases and in market operations.

Farmers have to become better managers

A farmer is no more the master of his farming, as he was in the past. The character of farming has changed drastically. Its knowledge base is externalized. Its production inputs come from far off sources. The farm products also most often have to find outside markets. With the entry of money economy into rural life, investments and returns have become a normal part of production planning in farming. At the same time, farming is being conducted in the face of several imponderables and many uncertainties. Failures and under-achievements have become a common feature in farming, as it moves into the commercial plane. Farmers frequently attribute this to the vagaries of the monsoon and the fluctuations in the market. But, more truly, this is also due to some marked deficiencies in management practices in farming.

Understandably, management of subsistence farming is one thing, and management of business farming is another thing, altogether. Indian farming is now inexorably moving into the commercial realm. At this stage, in every development approach, management has to be as significant an input as the technology input. Therefore, farmer's capacity building in farm management has to be reckoned as the core responsibility of extension education.

A close look at this goal will reveal that the capacity building task must begin with knowledge management. New technologies, in their relevance, have to be recognized, understood and accepted by the farmers, along with the skills required to use them properly in their individual situations. Technologies vary in their complexity. They cost something to the farmer, at least in terms of time and energy. They have to be utilized to their full potential both in crop production and in farm resource management.

Further, farmers today will have to deal with many externalities. Securing several inputs, each from a different source, and marketing the farm products to their own advantage in unfriendly markets, is not an easy job. It helps if the farmers have prior knowledge of each of these sources and has certain amount of bargaining capacity to deal with them. But, this is not the case, at present. As such, most often, it is the farmers that are found at the receiving end.

In addition, handling financial transactions in present day farming has become a necessary responsibility of farmers. The task of planning the credit requirement, securing the credit, properly utilizing the credit and repaying the credit is in itself a very complicated matter. This is not currently within the managerial capacity of many farmers. In recent times, credit management has become the most distressing part of the agriculture modernization process. Venturesome farmers at least need a basic understanding of the economic aspects of the commercial farming, as part of their managerial competence.

For all these reasons, present day farmers have to be initiated into the management domain, to be confident and effective in contemporary farming enterprises. Therefore, the extension system has to equip itself adequately to serve the changing farm sector, avoiding the needless suffering of the farmers on one hand, and minimizing the inefficiencies in development on the other.

Under the Indian conditions, women have been the unseen hands in farming for ages. They play a critical role both in decision making and in farm operations. It will surely be a case of all-round benefit if they are brought into the development mainstream, more effectively. The extension system has been sensitive this issue, all along. It seems that the conditions are now becoming more conducive to devote a greater extension attention to this strategic matter.

Conclusion

It is to be stated that the world around the Indian farmer has changed very drastically, and very rapidly. Compared to the farmer in the developed world, the Indian farmer is more unprepared for this changed environment. As such, he faces the challenges that no other farmer faced in the history, so far. He is now required to make more adjustments in his life and in his farming, more extensively and much more rapidly. Also, he lives amidst a vast multiplicity of information sources, today. But, still, the agriculture extension service can play a pivotal role in enabling him to make the required adjustments as smoothly and painlessly as possible. It can serve the farmer as a dependable communication bridge and as a meaningful educational service at this critical time. It is really a question of the kind of extension leadership that the country throws up, and the kind of vision that this leadership brings into focus, in the future.

References

Ameur, Charles (1994). Agricultural Extension: A step beyond the next step *World Bank Technical Paper, Number 247.* The World Bank, Washington D.C., U.S.A.

Anderson, Jock R, and Comelis de Haan, (1992). Public and Private Roles in Agricultural Development – *Proceedings of the Twelfth Agricultural Sector Symposium*. The International Bank for Reconstruction and Development. The World Bank, Washington D.C., U.S.A.

Census of India (2001). Registrar General, GOI, New Delhi.

Dhruvanarayana, V.V, and Ram Babu (1983). Estimation of Soil Erosion in India, *Jrl., Irrg., Drain. Engg.*, 109(4), pp 419-34.

Dwarakinath, R, (2000). *Extension Implications of a Changing Agriculture in India,* Institution of Agricultural Technologists, Bangalore.

Dwarakinath, R, (2001). *WTO: Indian Farmer Left High and Dry,* Institution of Agricultural Technologists, Bangalore.

Dwarakinath, R., Sethu Rao, M.K., and Hanumappa, P., (1994). *Guide to Sericulture Extension*, Central Silk Board, Bangalore.

Land Use Statistics (1998). New Delhi, Directorate of Economics and Statistics, GOI.

Macklin, M. (1992). "Agricultural Extension in India. *World Bank Technical Paper*, November 1990. The World Bank, Washington, D.C., U.S.A.

Paroda, R.S. (1998). *First International Agronomy Congress*, Indian Society of Agronomy, ICAR, New Delhi.

Reddy, M.S. (1998). *Five Decades of Water Resource Development In India",* Paper for the Indian Water Resources Day, Indian Water Resources Society, New Delhi.

State of Forest Report, (1999). *Manorama Yearbook 2003*, Malayala Manorama, Kottayam, Kerala.

Subba Rao, I.V. (1997). Food Security in Harmony with Nature, (Editor Dr.K. Shivashankar) *Third IFOAM-ASIA Scientific Conference and General Assembly*, held at UAS, Bangalore 1st – 4th December, 1997, p 57 – 61, Yield Improvement Mission, 2001, Karnataka Agriculture Commission, GOK, Bangalore.

5

CHANGES MADE AND NEEDED IN THE AGRICULTURAL EXTENSION SERVICE IN ISRAEL

FROM CO-FINANCING AND COMMERCIALIZATION TO PRIVATIZATION

Abraham Blum

The obvious

Extension systems have their own history, shaped by a range of socio-cultural and political factors, as well as the types of farming, the relations between growers, government and civic society, and many more influences. Therefore, we cannot hope for one extension system to be "the best" and universally adoptable. The history of agricultural extension is full of failures, which often started with the "transfer" of a system that succeeded in one place , to another, in which the socio-cultural conditions were too different. However, case studies can highlight experiences that induce in the reader thoughts about what might be useful (or not) also in another situation. Therefore, past and future changes in the Israeli extension system will be discussed with a review on the changes in agriculture, which made it necessary for extension to adapt to the changed situation.

Beginnings

During the British Mandate over Palestine (1920-1948), agricultural inspectors gave much attention to the identification of pests on plants, which growers brought to the regional offices, but extension as such did not have a high priority. In the Jewish settlements, members who had more agricultural knowledge than others, went from settlement to settlement, and lectured there in the evenings. Courses were held by growers' associations. These also published books and periodicals. Scientists in the research station of the Jewish Agency (the main development organization) saw it as their duty to hold meetings with the growers in their villages. Thus, a close relationship was forged between researchers, advisers and growers.

Two parallel extension services for different clients

After the establishment of the State of Israel, in 1948, a **Professional Center** was set up in the new Ministry of Agriculture. Its main task was to collect and transmit agricultural knowledge, in cooperation with the research stations. During the first three and a half years after independence, 700 000 new immigrants, mainly survivors of the holocaust and other refugees arrived in the new country. The bulk of new immigrants could be absorbed only on previously uncultivated land. 250 new *Moshav* (cooperative) villages were established, all based on agriculture, but the new immigrants had no farming background. Therefore, a new approach to extension was needed. **Volunteers with practical**

experience were recruited. Usually, a couple from an established *Moshav* came to live with the new immigrants in their new village. The husband worked as agricultural "guide" (a term that is still used for extension advisers), his wife did the same in home economics. Most of these guides had only 10-11 years of formal education. They learned about new crops and practices, which were suitable to the development areas, while on the job. Since that time, in-service training of at least one day a week has become a typical feature of the Israeli extension service. Within less than four years after independence, the population had doubled, and thanks to these new immigrant growers and their volunteer advisers, the food situation in the country changed from scarcity of agricultural produce to surpluses. By 1960, most "new" growers had bridged the agricultural knowledge gap and the two separate extension systems could be merged into a joint Extension Administration. A few years later, the Extension Administration and the Professional Service of the Ministry of Agriculture were fused into the **Extension and Professional Service** (acronymed **Shaham**).

Shaham

Although Shaham's advisers are employed by the state, they see themselves as professionals, whose first duty is to serve the interests of their clients, like doctors and lawyers. They are not expected to advance governmental policy goals, when these clash with growers' best interests. In the late 1980's, Shaham's independent regional offices and those of the Ministry of Agriculture, which deal with planning and statutory tasks, were fused for efficiency reasons. Shaham staff feared that they might have to engage in future in supervisory and regulating activities, which might destroy some of growers' trust. Their fear was unfounded.

Shaham has some peculiar features, which have to be seen against the social and agricultural backgrounds of Israel. Nowadays, growers are highly specialized, and so are all advisers. These are attached for administrative purposes to a regional office, but they can be found there only one half day per week or even in a fortnight. The rest of the time, advisers are expected to be in the field. However, advisers' main affiliation is to one of 15 national subject matter departments – covering the different agricultural branches and services (soil, plant protection, irrigation, mechanization and production economics) All advisers in a specific department meet regularly, at least once a month, to discuss professional questions. This is possible, because the country is small. Due to advisers' high specialization, a grower is sometimes advised by more than one adviser.

When a subject area becomes diversified, special subgroups are formed, which devote their attention to their area of specialization, under the guidance of a **Chief Professional Adviser (CPA).** Thus, in the plant protection or production economics departments, there would be CPAs for vegetables, fruits, flowers und so on, and in the vegetable department – CPAs for tomatoes and other important crops. Depending on the relative importance of each branch and of the need to specialize to this level, such interdisciplinary specialists have a national or a regional responsibility.

Changes in agriculture necessitate changes in the extension system

When the home market became saturated, in the 1960's, surpluses were sent abroad. The first attempts to export quickly taught the lesson that for export - the quality of the produce had to be improved .Farmers had to change from mixed farming to to specialized branches, and accordingly, advisers too had to **upgrade** their own knowledge. Shaham started to release advisers (without loss of salary) to study towards a B.Sc. degree in Agriculture. This was mainly done on a part-time basis, so that the advisers did not have to severe their links with their clients.

After some years, it became clear that an even more specialized level of knowledge was needed. Many advisers worked for a Master degree. Their research thesis has brought them nearer to research. Planning, directing and summing up applied field experiments became over time a considerable part of the advisers' work. The development of new technologies became often more central to their work than spreading knowledge. Today, half of the advisers, and practically all who join the service, hold at least an M. Sc. degree.

Shaham advisers usually work three days a week with growers. A fourth day is dedicated to field trials, and a fifth to formal or informal in-service training, e.g. attending a course, reading in the library, working with researchers towards practical solutions to the problems they encountered. Annually, the 200 extension advisers conduct, analyze and summarize 500 field trials. As always, exaggerations can be observed, and some growers have complained that their advisers have become too absorbed in their research contacts and have less time to come out to their farms. This problem worsened, when the mileage allowance of advisers was curtailed.

What motivates advisers?

Although the salary levels of advisers in the government service is not high, compared to what the same graduates could earn in the private sector, relatively few extension workers left Shaham, once they had attained tenure. While the position of tenured civil servant, looking forward to receive one day a government pension, might be attractive to some, the chances of moving into a higher salary level is not quite motivating. In practice, there are only two statutory positions, to which advisers could aspire: to become head of department or regional director. Both positions become available only when the incumbents retire, and that might take a long time. Furthermore, only one team member could get the appointment. The position of regional director is mainly an administrative task, and the increase in salary is relatively small.

Until the manpower of Shaham and the mobility of the advisers were curtailed, and its fate became uncertain, due to a change in governmental policy, the morale of advisers was very high. They did not count their working hours by the stopper. A number of factors contributed to this high motivation. Extension advisers always saw themselves as professionals and pioneers. They were highly esteemed by the farming community. **Social approval** became an important incentive. It worked especially well in a relatively small system, in which everyone knew everyone else. The other aspects of professionalism are the **opportunity for constant upgrading of knowledge**, formally or informally, and the freedom and encouragement to **develop new initiatives**. Such advisers, who could earn more in the private sector, prefer the freedom to initiate new ventures over the need to "create more income" to their company. The extension service helps outstanding advisers to find funds for professional visits abroad.

Another incentive, which is not dependent on funds, is the "position" of **Referent**. An adviser, even a relatively junior one, who has shown initiative, can become referent for a new crop or for a special problem. Referents are released from part of their routine duties (without any change in the statute book) in order to become **the** experts in a new area of specialization. Referents will read all about the new crop or topic, which they want to advance. They will meet with the researchers involved, visit any place in which relevant experiments are conducted, and they might even travel abroad for this purpose. This challenge gives a referent the feeling of being number one. If the crop will turn into a new and expanding agricultural branch, the referent might become a CPA. Should the idea prove to be unpractical, referents can go back to their former duties and take up another challenge.

Adaptation of extension activities to different farming systems

The different village forms in Israel, which exist side by side, made it necessary to adapt extension activities to their respective farming systems. Most agricultural land in Israel, with the exception of Arab villages and some older Jewish villages, is not privately owned. It belongs to the nation and is leased to growers at a very low rent for a period of 49 years, with automatic renewal of the lease, as long as the land is under cultivation. The main forms of Jewish agricultural settlement are the *Kibbutz*, often referred to as collective settlement, and the *Moshav* , also known as cooperative village. These two forms differ mainly in the amount of cooperation within the village, and this has changed dramatically during the last 20 years.

In the Kibbutz, production means are owned collectively and the land is cultivated as one large farm. Different agricultural branches are managed by Branch

Coordinators, who usually have a good professional background and considerable experience. Members of the *Kibbutz*, who have specialized in a given branch, take turns as coordinators. Thus, the extension adviser can work opposite a knowledgeable client, who typically would attend upgrading courses and often handle experiments. Furthermore, the adviser could meet with the whole branch team, which works as a collective unit.

The situation is different in the **Moshav,** where every family manages its farm unit. The need to compete on the export market forced also *Moshav* growers to specialize. Unless an urgent visit is needed, e.g. when an unusual plant pest is discovered, Shaham workers used to inform the *Moshav* secretariat, at which date they will visit the village, so that interested growers can attend. They usually met the adviser on one of the farms, whereupon the adviser might visit several farm units for specific purposes. This system has changed with the growth of commercialization. Now, advisers visit *Moshav* growers, when invited according to new rules.

Land in **Arab villages** is privately owned. During peak seasons, the extended family is ready to help. The technological progress and the rise of productivity in the Arab villages were probably the highest achievement of the Israeli extension system. This traditional agricultural sector was developed mainly through extension activities, without breaking the traditional, socio-cultural patterns. The main strategies were: (1) to begin with young, practically well trained advisers, who belonged to "good families" (Hamullahs) and were more "acceptable" than others, (2) to emphasize plant protection and to suggest new crops, which the old farmer did not know – both because they show quick results, (3) to set up small trial plots, which neighbours could see, (4) to gain the farmer's trust, so that the male advisers could work also directly with the women, who did much of the farm work, and (5) to foster the self-assurance of the Arab growers (Blum, 1988a,b). Today, these strategies are no longer needed, and one of the early Arab advisers has become the Deputy Director of Shaham.

Changes in the use of instructional technologies

The early tradition of spreading agricultural knowledge through **talks and publications** continued also after independence, but it had to be adapted to the needs and intellectual skills of the new immigrants. Some had a long tradition of reading in Hebrew (mainly religious texts) and therefore had access to written extension materials. Others came from a totally verbal culture, but had a remarkable potential to retain what was said during a talk. Advisers learned to relate to these groups according to what seemed to be the most effective way.

Talks were accompanied by visuals and later **audio-visuals**. In the early days, slides were very popular. Shaham set up a unit, which prepared slides and other instructional aids. This unit became soon the most dynamic part of the extension methods department.

Agricultural advice given by **teletext** was tried out, but was not well accepted. Agricultural news were broadcast on the **radio,** for 10 minutes, early in the morning, but this time slot was not popular with growers. The topics changed from day to day, according to the weight of the different agricultural branches. Actually, in urgent cases, advisers could reach their clients quicker by postcards, and later by e-mail, than through the radio. When **television** came up, a much more popular, weekly half-hour program was emitted, in the late afternoon. When the percentage of growers in the total population diminished, the agricultural television programs were discontinued, in spite of their popularity.

With technological advancements in the area of instructional aids, and especially filming, a big step forwards was possible, especially after **video cassettes** became available. Quite a number of advisers teamed up with Shaham's professional film maker to produce so far some 200 videos. These advisers were also the people who themselves used the videos as audio-visual aids. Over time, the quality of the videos grew, and several productions were rewarded with prizes at international agro-film festivals. Films were also sold to agricultural schools and private buyers.

The next change came with the appearance of the **DVD**. Now, films can be kept better on disks and can be controlled by a lap top computer. The advance of "power point" meant an end to the long reign of slides.

When growers started to buy home **computers**, the production economists of Shaham developed the "Mayan" farm management software. Later, program development was taken over by commercial companies, but Shaham continued to advise "Mayan" users. With the development of the microcomputer, new potentials for extension were foreseen (Ausher *et al.*, 1993), but nowadays, growers - both in Israel and abroad - seem to use computers mainly for communication purposes, e.g. **e-mail,** and less to retrieve new agricultural information (Gelb & Offer, 2005).

Among all the technological developments, none had apparently a greater effect on the contact between grower and his adviser than the **telephone**. It is quite accepted that growers call their adviser even at home, in the evening. Sixty-three percent of farmers actually used this phone contact (Blum & Azencot, 1991). Nowadays, when all growers and advisers have their **cell phone**, this has become the most commonly used means of communication between the two.

Group activities have increased, when advisers could no longer serve each grower individually. Growers' clubs exist mainly in the more intensive, horticultural branches. Some clubs were initiated by the regional adviser, who invites also speakers on specific topics, or by local growers. Growers' clubs meet either in the afternoon (e.g. in the less densely populated desert area) or in the evening (when little travel is involved).

Changes in extension training and research

During the 1960's and 1970's, Shaham's **Extension Division** was very active, training new advisers, organizing in-service workshops, developing (together with growers' organizations) packages of publications, which were regularly updated, and experimenting with new extension methodologies. Later, its output and prestige diminished (Kiron, 1989).

In the 1960's, Shaham approached the **Faculty of Agriculture** of the Hebrew University to set up **a graduate extension department**, which should train students to become advisers in the extension service. The first attempt failed after a short while, but a second try, towards the end of the 1970's, was more successful. The university provided the lecturers and researchers, and Shaham's senior extension advisers agreed that students should accompany them on their tour of duties. Students appreciated the opportunity to observe in practice what extension was all about. This second attempt at university based extension training lasted some twenty years. During that time most research on agricultural extension in Israel was conducted at the extension department of the Hebrew University. When fewer students enrolled in production agriculture, and the university had to cut down its academic manpower, the positions in agricultural education and extension were no longer filled, after the incumbents had retired. Thus, the department shrank, until its last professor became an emeritus.

Research - extension- farmer interactions

As already shown, extension advisers devote a considerable part of their time and efforts to **field trials**, either on regional experimental farms or (in a second stage) as controlled demonstration plots , on growers' fields. At the end of a season, researchers, advisers and interested growers meet at a national **end of season gathering** to discuss the experiences gained during the season with a specific crop or branch. Shaham prepares towards these meetings written abstracts of relevant experiments. The full texts are supplied on CDs. However, many growers do not wait for theend of season gatherings and approach researchers also during the season. Cinnobar (1993) found that 76% of growers had contacts with agricultural research. Another important institution, in which researchers, extension advisers and growers cooperate are the **Research and Development Authorities** in development regions (Blum, 1991).

Growers pay for a substantial part of research. For the last 20 years, they have been paying, through production and marketing boards, 0.625% or more of the value of their produce - for research and extension. This levy covers about 13% of the total annual agricultural research budget (not much less than the 15% contributed twenty years ago). This is probably a record figure, worldwide, for farmers' contribution to agricultural research. Ninety percent of the levy is earmarked for applied research and 10% for extension. The research fund is distributed to 16 branch directorates. These decide about the use of the funds. The deduction for marketed tomatoes, for instance, would go to the vegetable branch directorate and that for eggs to the poultry branch. Each branch directorate is composed of extension advisers, as representatives of the Ministry of Agriculture, researchers and growers. These representatives decide about the priority topics for research and call for research proposals. Usually, one of the advisers serves as chair person. The proposals are reviewed by smaller committees, which are also composed in the same tri-partite style. Thus, growers' representatives have a strong influence on agricultural research policy and priorities.

The farm situation changes drastically - with implications for extension

It is only natural that with growing efficiency in agriculture, fewer growers are needed to grow more. That happened also in Israel. In 1948, one grower produced food for 17 persons, now - for over 100. Agricultural output grew by a factor of 12 in real terms, and 30% of the produce is exported. However, with growing input costs and lower market prices, many growers can no longer make ends meet. Especially labor cost increased dramatically in the last 15 years, from 33% of growers' gross income to 67%. During the same period, the number of independent growers diminished by nearly two thirds! Parallel to this development, also the extension service shrank, though not at the same rate. In spite of the smaller ratio between growers and advisers, the latter could visit farmers less often during regular working hours, because each adviser had now to serve a larger area and travel allowances were severely cut. Mainly positions in the regions became vacant, changing the ratio between advisers in the regions and those in the center from 3:1, at the peak of Shaham's expansion, to 2:1 today. This again, made visits to farms more difficult. The frustration over their inability to serve farmers as well as before, and the rumors about impending privatization affected advisers' morale.

It is still difficult to fire tenured advisers, who have the status of government employees. Due to cuts in the number of advisory positions, only few young agronomists could join the service. Today, only 9% of the advisers are under 40 years old, and 56% are over 50. No wonder that these older advisers fear to be thrown into a free market, at an age it is difficult to find employment or start a business.

From co-financing to commercialization and privatization

In many countries, extension services started to differentiate between topics of general interest, which deserve free advice, and services, from which only one farmer profits, e.g. soil analysis. In Israel, Shaham's Irrigation and Soils Field Service was financed from its inception, some 40 years ago, jointly by the government and the Regional Councils. The latter levy taxes from growers.

When mileage allowances for advisers were cut, growers' organizations decided to fill some of these gaps. The first to suggest paying for "additional extension" were the avocado growers in one district. This enabled Shaham to employ one more avocado adviser. The advisers also got an additional mileage allowance and extra remuneration for their more intensive work (Elkana, 2001).

In the next step, Shaham advisers were allowed to work for extra pay after working hours or on weekends. This work should be outside their routine job. Growers are ready to pay advisers to supervise their fields, especially as part of their integrated pest management. Growers pay either directly to Shaham, or through a growers' association, and Shaham adds the money to advisers' salary. These payments account for about 17% of the total extension budget.

Such "additional extension packages" were introduced into most agricultural branches. This quasi privatized extension, sanctioned by the State, goes to the extreme in the citrus branch. Growers sign up on a yearly basis, based on their crop area and the requested visits. Growers, who do not sign up, receive annually only two farm visits The head of Shaham's Citrus Department pools the fees collected and dispenses them to the agents, taking in account the distances the advisers have to travel. In other departments, advisers receive the fee paid by the growers according to their reported overtime (which is checked by the involved growers' organization). According to a new arrangement between the Flower Board and the Floriculture Department of Shaham, the Board pays Shaham for each on-farm visits by an adviser about twenty US$ per hour. The flower board covers 75% of this sum from the marketing levy. The grower pays the other quarter. Each grower is entitled to select the extension adviser with whom he would like to work, but the latter can refuse to work with certain growers. Yearly contracts are signed, allowing up to 24 farm visits. Additional visits can be purchased from Shaham "unsubsidized", but still at a much lower rate than private advisers would have to charge in order to survive. Summarizing, Wolfson (2005) reports "a tremendous increase of agent to grower commitment", but also that only 54% of flower growers subscribed to the new scheme – in spite of the heavy subsidy. Before the introduction of the new arrangements, 71% of flower growers have received on-farm visits from extension advisers. Those, who cannot pay for extension, are often the first to be thrown out of competition.

In many countries, governments subsidize extension supplied by privates and commercial companies. Here we have an opposite case: Growers and their organizations subsidize the Ministry of Agriculture and its staff.

This quasi privatization also introduced inequality among advisers. Some of the flower advisers earned under the new system 100% more additional income than before. Other advisers earned not so much, and a few saw their income even going down (Wolfson, 2005). This must lead to tensions within a department that used to work as one team. Some advisers left Shaham, because they felt they became "field inspectors" and had no more time for professional development.

Shaham is not the only extension provider. Farmers who grow for the canning industry are actually obliged to follow the instructions of the industry's advisers. Cinnobar (1993) estimated that 130 commercial firms and 40 regional and national agricultural organizations offer advice. More than 75% of this additional extension was provided for free. Only 25 private advisers were active. Growers clearly preferred the governmental extension advisers over all others.

Quo vadis, extension?

In 1994, the Israeli Ministry of Agriculture decided to transform Shaham into a parastatal, commercial company. At that time, the resistance of the advisers and the parliamentary farm lobby was still strong, and the plan fell through. In 2005, the cabinet decided to privatize extension successively, and to leave in the Ministry of Agriculture only a small nucleus of some 50 advisers whose task would be:

- to collect and transmit professional knowledge to the new extension providers,
- to advise on public goods and topics of national importance like saving water, environmental quality – but through external providers,
- to engage in applied research, in cooperation with governmental and other research institutes,
- to monitor and supervise the agricultural extension provided to growers, and to supervise the apparatus that manages the transfer of governmental subsidies to the external extension providers,
- to give professional advice to different units in the Ministry of Agriculture, to research bodies and other professional units.

A joint committee, chaired by the directors general of the Ministries of Finance and Agriculture is supposed to work out a detailed plan, how to implement the decision to privatize extension during the next four years.

With privatization, some major problems can be expected:

1. Extension will become more expensive, and the number of growers, who use extension advice, will further diminish. The example from floriculture,

which does not stand alone, shows that even now, in the commercialization stage of official extension, a considerable percentage of farmers stop using extension. Experience shows that once extension is privatized, governments tend to curtail subsidies - especially when the farming lobby is weak. With knowledge being a major production factor, this development will affect agriculture negatively and more growers will have to stop farming.

2. The Dutch experience shows that with privatization, knowledge is no longer flowing freely. Those who have to buy knowledge, or develop it themselves will not want to give it on for free (Proost and Duijsing, 2002). Confronted with a drastically reduced budget, Ministry of Agricultural researchers in Israel have to compete for international grants. These are usually allotted to more basic research issues. The outcome of that research needs further, practically adapted, technological development.
3. One of the major achievements of Shaham was its key role in technology development, through adaptive field tests. There is a real danger that the planned "nucleus" will be largely occupied by bureaucratic supervision tasks, checking private extension advisers' bills and requests for subsidies. The remaining advisers will have less time to visit farms and thus will be less knowledgeable about farmers' problems. This, in turn, will make it more difficult for them to give useful feedback to those in the Ministry of Agriculture, who need that information.

Rivera and Alex (2004) emphasize that governments are responsible to ensure the quality of a (privatized) extension system, and to provide extension on topics of general interest. The Israeli government's decision accepts these obligations at least by token, but it is quite unclear, how the "nucleus" of advisers can fulfill all the tasks mentioned in the government's decision. In the meantime, Shaham tries to postpone the implementation of the decision, and the Ministry of Finance threatens to cut the budget. So far, none of the stakeholders has come up with a realistic, operational plan - how extension should function in future. Are production boards or growers' organizations ready and able to take up the challenge? Who will be the driving force for adaptive research and technology development? Will planning companies expand into the extension business? Will ex-Shaham advisers and experienced farmers set up shop as private advisers? Will the future of extension in the country be planned or left completely to "market forces"?

During its long history, the Israeli extension system has managed to adapt to changes in agricultural conditions. Now, it stands before its greatest challenge. It is still not yet clear, how it will adapt this time.

References

Ausher, R., Blum, A., Gelb, E. and Marom, D. (1993). *The Potentials of Microcomputers in Support of Agricultural Extension, Education and Training.* Rome: F.A.O.

Blum, A. (1988a). Extension strategies used to develop a traditional farming sector in an advanced agricultural surrounding – The case of the Nazareth region in Israel. *Quarterly Journal of International Agriculture,* 27 (3/4): 299-313.

Blum, A. (1988b). Use of extension by traditional Arab growers in Israel and in the Gaza strip. *Quarterly Journal of International Agriculture,* 27 (3/4): 314-326.

Blum, A. (1991). An improved agricultural knowledge system: The Israeli experience with regional research and development authorities. In: W.M. Rivera & D.J. Gustafson (eds) *Agricultural Extension: Worldwide Institutional Evolution and Forces for Change.* Amsterdam, Elsevier, pp.139-150.

Blum, A. & Azencot, M. (1991). The contact between agricultural extension and family growers in Istael – with some international comparisons. *Communications* 16 (2): 251-261.

Cinnobar Advisers Ltd. (1993). *Survey of the Needs for Agricultural Extension.* Givataim: Cinnobar (Hebrew).

Elkana, Y. (2001) Extension Services in Israel. Paper prepared for *INRULED-UNESCO Workshop on "Technical Training for Rural Development: Looking to the 21st Century"*, Baoding, China.

Gelb, E. and Offer, A. (2005). (eds.) *ICT in Agriculture: Perspectives of Technological Innovation.* http//:departments.agri.huji.ac.il/economics/gelb-main.

Kiron, M. (1989). *Review of the Extension Division*, Tel Aviv: Shaham (Hebrew).

Proost, J. and Duijsings, P. (2002). The Netherlands: Going Dutch in extension, 10 years of experiences with privatized extension. In: Rivera, W.M. and Zijp, W. (eds.) *Contracting for Agricultural Extension – International Case Studies and Emerging Practices.* Wallingford and New York: CABI Publishing.

Rivera, W.M. and Alex, G. (2004). The continuing role of government in pluralistic extension systems. *Journal of International Agricultural and Extension Education* 11 (3): 41-52.

Wolfson, D. (2005). Recent developments in financing the floricultural department of the Israeli Extension Service. Perth: *Acta Horticulturae* 672 (IV. International Symposium on Horticultural Education, Extension and Training).

6

The Rise and Fall of Training and Visit (T&V) Extension: An Asian Mini-drama with an African Epilogue

Jock R. Anderson, Gershon Feder and Sushma Ganguly

Introduction

Within the voluminous literature on agricultural extension experiences, the emergence, widespread diffusion, and ultimate disappearance of the Training and Visit extension system has received much attention. Various country-specific or generalized discussions were contributed over the years. The purpose of this chapter is to pull together these insights and observations so as to provide an analytical narrative of the dramatic international rise and fall of T&V extension. To tell a coherent story requires utilizing a conceptual framework explaining the organizational behavior of public extension systems and the incentives of their staff and management, as well as the behavior and incentives of external donor agencies. Thus, following the next section, which provides a general description of the pre-T&V extension scene, we sketch the general behavioral patterns underlying our discussion. This is followed by a brief description of the principles of the T&V system and their rationale and merits, as well as their likely deficiencies. The subsequent section of the chapter describes the evolution of the system in Asia and Africa (there was no significant presence of T&V extension in Latin America). The final section provides concluding remarks that attempt to generalize the insights gained. In view of the dominant role of the World Bank in the development and spread of T&V extension, there will be heavy reliance on the actions, studies and deliberations of the World Bank, although other sources are utilized.

Extension in the Post-Colonial Era

In the immediate aftermath of the dismantling of colonialism, agricultural extension organizations in developing countries followed generally the administrative structures of the colonial era (Axinn and Thorat, 1972). The main focus was on supporting the production and marketing of export crops. Gradually, the newly independent countries shifted more attention to food production, and their extension services evolved to serve a large clientele of smaller farmers rather than the export crop estates. The production orientation of earlier years was often modified to include more general support to various aspects of farmers' activities, such as community work, credit, input supplies, nutrition, family planning (e.g., Anderson and Hoff, 1993). A common perception among policymakers and

---

The findings, interpretations, and conclusions expressed in this paper are entirely those of the authors. They do not necessarily represent the views of the International Bank for Reconstruction and Development / World Bank and its affiliated organizations, or those of the Executive Directors of the World Bank or the governments they represent.

agricultural officials in those times was that farm productivity growth was constrained by farmer backwardness, inadequate organization, and deficient local leadership. Accordingly, in the 1950s and 60s, extension agents operated within multipurpose development programs and village development initiatives. In many instances, they were administratively assigned to local governments or to central ministries other than agriculture, or were accountable to more than one agency (Picciotto and Anderson, 1997).

The next section explains the factors underlying the observed performance of public extension and the political and bureaucratic environment in which they operated. This will provide the background for the emergence of the T&V extension model, which aimed at overcoming some of the prevalent deficiencies.

A Characterization of Extension Organization Behavior

Many scholars and observers of rural development commented on the frequent manifestations of unsatisfactory extension performance (e.g., Rivera Qamar, and Crowder, 2001). The common incidence of such performance issues across many developing countries is highlighted by Feder, Willett and Zijp (2001), who identified eight interrelated characteristics that are inherent in public extension systems, and which jointly lead to inadequate performance. These characteristics tend to be generic and could be applicable to any sector/subsector, but we highlight below why they have been more binding constraints in the case of the extension programs.

1. **Scale and complexity:** The cost of reaching large, geographically dispersed and remote smallholder

farmers is high, particularly given high levels of illiteracy, limited access to mass media, and high transport costs. Farming systems often entail several crops, livestock, and even within given geographical area, there are variations in soil, elevation, microclimate and farmers' capabilities and access to resources. With such a large and diversified clientele, only a small fraction of farmers can be served directly (face-to-face) by extension, and agents tend to focus on the larger, better resourced and more innovative farmers. This reduces the potential for farmer-to-farmer diffusion. Limited outreach to smallholder clientele reduces the benefits and impact of extension.

2. **Dependence on broader policy environment:** The outcome of extension efforts depends on policies over which agents and their managers have little influence (input and output prices, credit policies, input supplies, marketing and infrastructure system). The policy environment can indeed affect any investment, but more so for extension programs, particularly if the advice is not adjusted to reflect the incentive environment in which farmers operate.
3. **Interaction with knowledge generation: *Public*** extension and research systems often compete for budget, but research institutions have an advantage due to their higher status, better management quality, and links with the global science community. This creates tensions and militates against an effective two-way communication (e.g., Mureithi and Anderson, 2000). The dependence of extension programs on science and technology (and generation thereof) and vice-versa is strong, i.e., the linkage effect is more important than it is among other sub-sectors, and yet research scientists do not have strong incentives to interact with extension.

4. **Difficulty in attributing impact:** Attribution of impact is an analytically challenging task in the case of extension programs, due to the lack of baseline information, unavailability of appropriate control groups, and the systemic biases in extension placement and contacts (Birkhaeuser, Evenson and Feder, 1991). Given the difficulty of attributing outcomes, managers focus on collecting the reporting input indicators (visits made, meetings and field days arranged, sample kits distributed), that are easier to assemble, and the incentives to deliver high quality information service is weak (Martens et al., 2002).
5. **Weak accountability:** Weak accountability (linked to the inability to attribute impact) is reflected in low-quality and repetitive advice given to farmers, and in diminished effort to interact with farmers, and to learn from their experience. There are no mechanisms, or incentives to create mechanisms (e.g., participatory processes) for fostering accountability to clients who can best observe the quality and quantity of extension input.
6. **Weak political commitment and support:** Feder, Willett and Zijp (2001) suggest that highly visible irrigation or road projects are more attractive to politicians than extension expenditures. This, compounded by difficulties in attribution of impact, makes extension an easy target for lower budgetary support.
7. **Public duties other than knowledge transfer:** Governments are often tempted to use field-level cadres of public servants who are already present in the countryside for non-extension duties such as collection of statistics, distribution of subsidized inputs, assisting and collecting loan applications, and election campaign work on behalf of local or national ruling parties.

8. **Financial unsustainability:** This can cause the demise of any investment program but it is observed that at times of general fiscal constraints, extension budgets are more likely to be squeezed due to weak political support. Due to the large share of staff costs in the recurrent costs, and the rigidity of public personnel policies, field operations, (including maintenance of equipment and buildings, and training) provide the most flexible cost-cutting budget items.

While the discussion above provides rationale for the weak performance of extension, and the limited popularity of extension spending with national politicians and senior officials, one observes a much greater enthusiasm and support for funding extension among managers and staff in external donor agencies and nongovernmental organizations (NGOs). The NGOs typically operate at a small scale (tailored to the limited external budgets they can provide) and their activities are on a full grant basis, not requiring local government funding. They are therefore less concerned with issues of fiscal sustainability. They overcome the various incentive problems by deploying external or local advisors who work closely with local extension at the field level, thus providing effective supervision. And they often have access to better quality, more committed field advisors, due to the higher pay levels they provide. Thus, NGO-supported extension initiatives often are perceived to (and may in fact) have better performance, but they cannot be realistically scaled up to national level with the same level of staffing and performance. Small bilateral donors often operate in a similar mode to that of as NGOs.

The incentives and explanations are different for large external donor agencies, which cannot have extensive field presence in the large extension projects they finance. An explanation of their support for extension projects, focusing on extension systems at the national level (or large sub national units) is provided by Anderson and Feder (2004). Often a small pilot or a small component within another project, with heavy supervision or other conducive circumstances (e.g., irrigation development, or delivery of abundant subsidized inputs, or simply the small and easily-managed scale) will create a perception (often justified) of success. The extension model of the small-scale pilot will then be promoted to both the donor agency management and to developing country policymakers as worthy of scaling up to the national level. The traditional reluctance of national policymakers is temporarily overcome by the availability of abundant external funds that are provided outside of the normal budget framework. Ministry of Finance officials, who normally are averse to the implied long-term liabilities entailed with expanding staff-intensive systems, do not know enough about agriculture and extension, and may be impressed with the output growth projections based on extrapolations from the small-scale pilots. Local extension and agricultural bureaucracies are supportive of the expansion as the external funding provides resources to improve their infrastructure, expand their organizations, and boost their status. The implicit rivalry with the research system is often overcome when the foreign-financed program focuses on both research and extension, or apportions some external funding for research organization-executed activities within a national extension project.

From the perspective of a large external donor, the upscaling of a pilot extension model to the national level is relatively simple to design, involving only one counterpart agency (the Ministry of Agriculture, or a couple of specialized agencies within it), thus reducing bureaucratic complexity. The activities funded within the project are well defined inputs: construction and refurbishing of extension offices and staff housing, procurement of vehicles, motorcycles (even bicycles), training activities, funding of additional personnel, etc. A national-scale project entails a large loan or grant, while the preparation cost to the donor is not proportionately as high due to the ability to extrapolate from the parameters derived from the pilot project. The cost of project preparation per dollar granted or lent is thus not high. Furthermore, repeated projects are easy to process. Extension projects do not present a hurdle as they do not require a rigorous ex-ante cost-benefit analysis (which confronts hardware-type projects). Moreover, the difficulty of attributing impact implies that it will be almost impossible to establish a definite failure at the end of the implementation period. Thus, the bureaucratic risk for the promoter of the project is only the possibility of flagrant non-implementation of the input side of the project (e.g., non-filling of agreed staffing levels, or no construction of agreed structures, or curtailing of field operations due to inadequate recurrent budget transfers from government). These are relatively low risks during the external-funded phase of an extension program and, even if they materialize, the cause of the failure can be assigned to the government rather than to a faulty design concept or to insufficient verification of the availability of viable technology to disseminate.

There are, therefore, strong incentives for external donor staff to promote large extension programs, and the local lack of enthusiasm can be overcome temporarily, for a couple of project cycles (10-12 years). Ultimately, senior donor officials expect extension activities to be absorbed in the normal budget process of the recipient country, as the external funding is viewed by them as a one-time investment cost. But if the expanded extension system entails a much larger recurrent (as well as equipment replacement) cost than that which is normally politically supported, a moment of truth is eventually faced. The local budgets will be drastically or gradually reduced, with a major negative impact to be observed in field operations (i.e., diminished interactions with farmers), as staff levels can only slowly be reduced. Alternatively, extension staff will be diverted into other (non-extension) agricultural or community activities for which budgetary resources are available. Extension size, or the volume of extension information dissemination activities, will thus converge to those dimensions that are compatible with political realities (e.g., Purcell and Anderson, 1997).

This projected scenario may span a somewhat longer time period in heavily donor-dependent, very poor, countries. The reason is that, in such countries, a significant proportion of government budget is derived from donor funding directly or indirectly, and recurrent costs are routinely funded (at least indirectly) to a significant extent by external donors. The concept of "mainstreaming" (i.e., absorbing into the normal budget process) of a larger extension organization is therefore less applicable, and senior donor management can be persuaded to support extension funding over a longer horizon. Furthermore, local policymakers are more receptive to what is perceived as donors' strong advice due to their dependence on external funding, thus misgivings about the priority given to extension are often not voiced too forcefully. One would thus expect that large donor-supported extension systems will tend to last longer in donor-dependent Africa than in Asia.

The Advent of Training and Visit Extension

In the late 60 and early 70s, technology diffusion became the key focus of agricultural extension systems. The rationale for shift was eminently apparent: new high yielding varieties for staple cereals were released by international research centers, and adapted by national research systems. This coincided with higher output prices due to food scarcity making dissemination efforts likely to meet receptive farmers (Lipton with Longhurst, 1989). Yet, extension systems of the time were viewed as not very effective.

The concept of Training and Visit (T&V) extension was developed in the early 70s, and implemented as a component in two regional irrigation projects, the Seyhan (Phase 2) project in Turkey, and the Chambal (Rajasthan and Madhya Pradesh) project in India, both funded by the World Bank in 1974. The principles of the system were spelled in great detail in Benor and Harrison (1977) and Benor and Baxter (1984), and are well familiar to most rural development scholars. We recount them briefly with an interpretation demonstrating that they were designed to overcome some of the inherent weaknesses of public extension systems outlined in Section III.

(i) A hierarchical organization with several layers of management overseeing a large cadre of village level workers under a single line of command, so that extension workers are not controlled by, or responsible to, other authorities (such as communal leadership or special crop-oriented organizations). The organization included subject matter specialists who were the technical resource staff.

(ii) A rigid bi-weekly schedule of visits to a defined fixed list of contact farmers (later modified to contact groups) in specific villages within a village worker's area of responsibility. The contact farmers (or groups) were expected to disseminate information further to other farmers within the community.

(iii) Fortnightly regular training of village level workers, administered by superiors and subject matter specialists, and focused on the information to be delivered in the coming couple of weeks.

(iv) No involvement of the extension organization and its field level workers in non-extension duties, such as input distribution and loan applications. Extension staff was to handle only agricultural information services.

(v) Regular interactions (through seasonal workshops) of extension's leaders and subject matter specialists with research station scientists. Specialists were expected to conduct their own on-farm research.

(vi) Concentration on the most important crops, and on messages about relatively simple low-cost improved practices.

The hierarchical structure, and the strict schedule of visits to predetermined contact farmers, whose names and time of visits were known to supervisors, was designed to overcome the weak incentives to interact with farmers on information delivery. The single line of command and prohibition on non-extension duties was intended to overcome the temptation to assign (by higher level officials) and undertake (by extension workers) potentially remunerative duties not involving information dissemination. Thus, the design improved the accountability of village-level workers to their superiors with respect to the quantity of their work, but not with respect to quality. In the absence of accountability to farmers, quality and impact were still difficult to observe, although the frequent and mandatory training was intended to improve the overall knowledge of workers and overcome their disincentives to invest in improving their skills. Feedback from farmers through village workers to supervisors and subject matter specialists was theoretically feasible during the training sessions. But the incentives for staff to solicit such feedback and report it were not much changed, compared to the past, in the absence of accountability to farmers. Some T&V projects (notably in India) had built-in units for monitoring and evaluation (M&E) that were designed to provide management with feedback, both on the mechanics of the system (visits, training sessions), as well as on impact, but the typical skill profiles of those likely to serve in such units were not adequate to perform reliable attribution analysis. Furthermore, these M&E units were structured to be under the authority of extension management, so their independence was compromised. As a result, the inability to attribute the extension system's impact convincingly at the aggregate level, and thereby to gain the support of policymakers, was still a problem under the T&V design.

The pre-designed interaction with research, and the structured positions for subject-matter specialists, whose mandate required regular contact with research, were to overcome the weakness in this field. But the incentives for research scientists to invest heavily in interaction with extension have generally been slow to change if at all, so there were limits to the progress that could be achieved in this respect.

The large clientele was to be served by a larger field-level cadre than before. This, as well as many other features of the design (multi-level hierarchy with more mid-level managers and technical experts, better mobility through vehicles and motorcycles, more local-level offices) implied not only a higher investment cost, but also a much higher recurrent cost compared to the pre-T&V situation. This is reflected in the fact that 90 percent of completed T&V extension projects reviewed in Purcell and Anderson (1997) encountered serious recurrent cost funding shortages, some even while the project was still under implementation. In our judgment this is the most crucial aspect responsible for the ultimate fall of the T&V extension model within a 25-year period from its inception in 1974, rather than the fact that the design contained features that limited the improvement in effectiveness. The promoters of T&V, while being aware of the cost, did not appreciate the political ramifications of high recurrent costs. Thus, Development Communication Report (1978, p. 9) cites a key leader of the T&V concept in response to a question on the high cost of the system: "cost is scarcely a consideration. No country in the world can afford not to spend this money".

The deficiencies in effectiveness (in the many places where such was the situation) could not be easily detected and linked rigorously to poor outcomes at the farm level, and it is unlikely to have caused by itself a disenchantment with T&V by agricultural policymakers. It is thus the higher cost, combined with the lack of convincing evidence of major gains attributable to extension, which most probably induced the fall of T&V.

The Spread and Decline of T&V Extension

Following the formal introduction of T&V extension components in the Chambal Irrigation Command (Rajasthan and Madhya Pradesh states in India) and the Seyhan Irrigation project in Turkey in 1974, World Bank staff associated with the projects provided enthusiastic feedback on the contribution of extension in these projects, and attributed significant on-farm gains to the extension reforms. The on-farm changes in these project areas have never been analyzed rigorously. It is quite likely that favorable farm performance was indeed observed in the project areas. But it as likely that such improvements were due in some part to the better management of the irrigation services, more timely availability of inputs, and the close attention of World Bank staff in a relatively confined project area (Moore, 1984). Furthermore, it is easier to manage a relatively small extension staff in an irrigation command area, operating within a fairly homogenous farming system. But concerns regarding the legitimacy of extrapolating from these apparently successful smaller scale projects to the state or national level did not seem to have been raised forcefully.

Thus, the perceived success induced World Bank staff and management to actively promote the adoption of the T&V extension model in India and other countries even before the completion of the two irrigation projects. This campaign, backed by readily available funding from the World Bank, was undertaken with such zeal and dedication that some observers likened it to the fervor of a religious movement (Moore, 1984, p. 307). The spread of the new concept was dramatic: By the end of 1982, the system was introduced in 13 states of India (with four more states in the next few years). In the same period (1975-1982), T&V extension was introduced with World Bank funding in ten additional countries (mostly in Asia and Africa) through self-standing extension projects, and in nine additional countries as components of larger agricultural projects. Other donors, such as the International Fund for Agricultural Development (IFAD) and the FAO adopted the approach (or aspects of it) in their operations as well. Later in the 80s, additional countries adopted the approach, and many of the projects that were initiated in the 70s and early 80s were continued or expanded (with new funding from the World Bank) in second-phase operations. Altogether, close to fifty developing countries utilized some form of T&V extension during the period 1974-1999. However, as second-phase projects in India and other large Asian countries were being completed, in the late 80s and early 90s, there were no new foreign-funded programs, as the external donors expected that these systems were already "mainstreamed". Thus, the extension systems had to be maintained through national or state budget only. In Africa, however, there were still new or follow-up projects through the 90s with the last one funded by the World Bank in 1998 in Cameroon.

It was inevitable that such a global effort attracted critical attention of development scholars and practitioners. Critical comment focused on both systemic areas of weakness of public extension systems, as well as on features that were more specific to the T&V design and the manner in which it was implemented. The earliest significant criticism is in the collection edited by Howell (1982a), which engendered a heated debate (Howell, 1982b), and was further elaborated by Howell (1983). The next significant forum in which serious concerns regarding the effectiveness of the T&V model were prominently highlighted was a 1983 workshop in the National Institute for Rural Development in Hyderabad, India. These concerns, pertaining to the experience in several Indian states, were summarized by Moore (1984). Some of the problems related to aspects of T&V that could not be fully implemented. Moore recounts reports of cases where the regular training sessions were not being held, or lacked real content; officers appointed as subject matter specialists without having adequate qualifications; linkages with research not quite enhanced; village agents not following the regular visit schedule, or contact farmers not attending visits, with some of them not being aware that they were designated as contact farmers. The common preference of agents to interact with larger scale and richer farmers was said to still be present under the T&V system. And insufficient attention to the supply conditions of inputs hampered the relevance of the information conveyed to farmers.

According to Moore (1984), the supervisory staff did not have incentives to use the strict visit schedule as a device to enforce work delivery by village workers, and at a more senior level, the interest in the T&V program was not genuine, reflecting merely the desire to obtain the enhanced resources (e.g., vehicles, offices) associated with the projects. The high cost of the system was Moore's key critical point.

Criticism and debates of this nature would have had a more decisive impact in forcing greater selectivity and flexibility in the global introduction and implementation of T&V in the 1980s had they been backed by systemic data sets confirming the checkered adherence to T&V design principles, and a larger number of rigorous impact studies assessing the farm-level results of the program. But extensive data on the mechanics of T&V in India were only available from the extension M&E units, and these indicated reasonable (albeit far from perfect) adherence to visit and training frequencies and contact farmer familiarity with their roles (Feder et al., 1986, Feder and Slade, 1986a, Slade et al., 1998). It was argued by some observers (e.g., Moore (1984), Chambers (1992)) that the M&E units' reports were likely distorted to present a favorable picture, but the extent of the bias was difficult to establish. An independent data set from one district in Haryana demonstrated, for that particular area, that the mechanics of extension operations were reasonably in place (Feder and Slade, 1986b). A subsequent critical report by Axinn (1988) reiterated some of the points raised by Moore (1984), based on extensive field visits, but rigorous impact studies were not available for quite some time. One such study, from one district in Haryana, indicated (under the strong assumption that benefits observed in the first three years of the project will continue in future years, although at a declining rate) that the incremental cost of T&V in that district was justified, as there was a high likelihood of a viable rate of return (Feder, Lau, and Slade, 1987). But there were no other rigorous impact studies until the 1994, when Hussain, Byerlee and Heisey found no T&V impact in Pakistan. A 1993 study by Bindlish and Evenson reported large returns for T&V extension in Kenya, but it was contradicted by Gautam (2000), who pointed out some data deficiencies in the earlier study, and demonstrated, using better data, that the impact in Kenya was insignificant. These studies have not played much of a role in the apparent lack of support in Asian countries for sustaining T&V in the late 1980s and early 90s. These countries (such as India, Bangladesh, Indonesia, Pakistan), have reduced the extent of their T&V extension operation over a relatively short span of time, both by not replacing retirements and transfers, by adding other duties or reassigning staff to other roles, and by abandoning key aspects of the T&V concept (e.g., the strict schedule of visits and trainings, and the uniform ratios of field and supervisory ranks).

The rearguard battle of the T&V extension concept was fought out in Sub-Saharan Africa, with the promoters facing increasing doubts and criticisms not only from rural development scholars and officials in developing countries, but also from within the World Bank itself. As early as 1985, the World Bank's Operations Evaluation Department expressed some criticism (albeit muffled) of the wholesale application of large-scale extension projects in situations that may not have warranted such systems (OED, 1985). The internal debate intensified in the late 80s and early 90s as it became apparent that Asian countries, and in particular India, are gradually withdrawing from the T&V model due to the high cost. This debate is evident in a World Bank 1989 edited volume on extension in Africa (Roberts, 1989), where several papers raise critical issues pertaining to the T & V approach's design and cost. The editor, in one of the papers, noted that promotion of the approach universally with little sensitivity to needs in particular situations and countries is attributable in part to the over-enthusiasm and inexperience of World Bank staff impressed by the initial favourable feedback from Asia. Thus, the author conceded that, in the design of some African T&V projects, there have been instances of unnecessary rigidity (Roberts, 1989, p.23). The paper also recognized that T&V designers had not done justice to the issues of true farmer participation and to the financing of the high recurrent costs.

However a 1990 report outlining lessons and suggestions for extension strategies (World Bank, 1990) refrained from explicit criticism (earlier drafts were more critical) and highlighted the greater flexibility and adaptability of the T&V design as applied in recent projects. Flexibility was indeed forced by reality on the originally rather rigid camp of T&V extension promoters. Examples of modifications in the design were cited by a number of observers : Roberts (1989) pointed out that in Sudan extension agents continued also their duties as supervisors for the irrigation system. Similarly, in the Comoros extension agents handled input supply functions as there was no reliable private supply system. The involvement of extension staff in non-extension duties was generally frowned upon by T&V designers. Similar deviations from the exclusive devotion to extension are mentioned by Bagchee (1994). The latter author also pointed out reduction in the frequency of staff training sessions in Kenya, Tanzania and Zaire, and higher ratios of farmers to agents in many countries compared to the standard 1 per 800 favoured by T&V designers. Increased use of mass media as a complementary channel of technology diffusion, along with the T&V effort, took place in Cote d'Ivoire, Kenya,

Nigeria, Senegal and other countries. The original T&V concept underplayed the reliance on mass media. A fairly common adaptation in African T&V countries was the shift away from "contact farmers" to "contact groups" as the foci for interactions between agents and the farming clientele (World Bank 1990, p. 21; Bagchee, 1994). This was designed to increase the number of farmers receiving face-to-face service from extension in view of the limited diffusion from contact farmers.

The modifications did not, however, tackle significantly a key source of the problems, namely, the staff-intensive high recurrent cost structure. High-cost issues were indeed highlighted in Bagchee (1994, e.g.,p.19). The conclusion was, however, only to exercise caution in expanding systems. By the time an influential World Bank Operations Evaluation Study was published (Purcell and Anderson, 1997), it was apparent that fiscal issues were the key challenge, and the various modifications in design offered (and experimented with) in the World Bank's African T&V extension projects were not sufficient to reduce the costs, as they came after the extension systems were already enhanced and expanded significantly. The flexibility in retrenching public sector employees was limited and thus staff costs could not be changed much. The 1997 OED report listed a number of frequently occurring difficulties in T&V extension projects and concluded that the relatively high (70%) share of projects graded "satisfactory" was probably not a realistic reflection of the actual rate of success due to the imprecise nature of such summary judgments. The study concluded that "the Bank had erred in the extent to which it has promoted the T&V extension management system in relatively uniform packages of investments and extension practices in large state and national programmes." The authors further concluded that the increased flexibility in African T&V

projects was not likely to resolve the problems encountered and therefore the T&V design was "unlikely to be the most appropriate approach for improving extension in many African countries" (p. 98).

The 90s witnessed greater interest in alternative extension concepts, with stronger participatory aspects, greater pluralism, and smaller public organizations. In Africa, in particular, there was donor pressure to transfer some aspects of extension to private or other non-government providers (Gemo, Eicher and Teclemariam, 2005). The World Bank's senior management was not inclined anymore to expend political capital in convincing reluctant leaders in client countries to accept a high-cost extension model for which there was no rigorous evidence of widespread success at the farm level. Furthermore, with a significant camp of critics inside and outside the Bank becoming increasingly vociferous in challenging the validity of the concepts, there were great risks in pushing further the T&V concept. It was thus in Africa that the T&V extension model breathed its last, ending twenty-five years in which agricultural extension received the highest level of attention it ever attracted on the rural development agenda.

Conclusion

As in most fields of human endeavor, fads are a fact of life. Development practitioners need to be aware that they are not immune from this aspect of the human condition and thus need to guard against falling ready victims to what may appear as a new and relevant approach to an old problem. A usually good starting point to assessing the new ideas is to check back: on the principles underlying them; on the experience in implementation; and on the lessons learned, from the past fads that were followed. Our discussion of the experiences, largely connected to World Bank operations, with the T&V approach to extension from birth to death, has been intended to facilitate such checking back. Several insights emerge from these reviews, which have a bearing on other current or future extension concepts, and more generally on rural development initiatives.

Approaches that have a better chance of being successful are those that highlight the intended beneficiaries as empowered clients with their needs and demands articulated through service arrangements that heed them, and do so in organizational structures that are affordable. Such arrangements work best when the empowerment is driven by personal client stakeholding in the delivery of the services being sought, as the client is best positioned to observe the utility of the services rendered.

The T&V extension experience also demonstrates the importance of timely and careful evaluation of pilot experiments with realistic assessment of the likely challenges to be faced if seemingly promising innovations are scaled up significantly. Of particular importance in this context is an analysis of the fiscal implications of the expanded scale, the degree of dependence on external funding, and the likelihood of domestic support to the recurrent costs of the expanded program over time. The often observed inadequacy of operations and maintenance funds in infrastructure projects suggests that this is a common issue, not specific to extension.

Once innovative designs are implemented on a large scale, evaluative studies in a reasonable number of differing locations are crucial in order to form reliable (preferably statistically valid) and convincing judgments on the applicability and limitations of the idea being promoted in different social, economic, political, and physical environments. The availability of appropriate baselines and control groups is often essential for reliable evaluation, but these require prior planning, because their absence is difficult to rectify ex-post.

Evaluation studies need to be independent of those who have a direct stake in the success of the endeavor, with evaluation teams in which there are no preconceived positions on the innovation being reviewed. While most sponsors of evaluative studies are aware of the need for evaluators' independence, one encounters too often evaluation teams purposefully selected by the concerned stakeholders so as to include scholars who have already expressed themselves favorably on the idea being studied, and this can compromise objectivity. In situations where attribution of impact is analytically difficult, a full evaluation will require quantitative work by qualified professionals to complement qualitative assessments. In many cases such econometric skills will be drawn from academia and research institutes, where they more commonly reside.

References

Anderson, J.R. and G. Feder (2004). Agricultural Extension: Good Intentions and Hard Realities, *World Bank Research Observer* 19(1): 41-60.

Anderson, J.R. and K. Hoff (1993). Technological Change, Imperfect Markets, and Agricultural Extension: An Overview, In K. Hoff, A. Braverman and J.E. Stiglitz, eds., *The Economics of Rural Organization: Theory, Practice, and Policy.* Oxford University Press for the World Bank, New York.

Axinn, G. (1988). *Independent Assessment of Agricultural Extension Services in India*, Unpublished, Michigan State University, East Lansing.

Axinn, G.H. and S. Thorat (1972). *Modernizing World Agriculture: A Comparative Study of Agricultural Extension Education Systems.* Oxford & IBH, New Delhi.

Bagchee, A. (1994). Agricultural Extension in Africa. World Bank Discussion papers 231, World Bank, Washingrton, D.C.

Benor, D. and J.Q. Harrison (1977). *Agricultural Extension: The Training and Visit System.* Washington, D.C.: World Bank.

Benor, D. and M. Baxter (1984). *Training and Visit Extension.* Washington, D.C., World Bank.

Bindlish, V. and R. Evenson (1993). *Evaluation of the Performance of T&V Extension in Kenya.* World Bank Technical Paper 208. Washington, D.C.

Birkhaeuser, D., R.E. Evenson and G. Feder (1991). The Economic Impact of Agricultural Extension: A Review, *Economic Development and Cultural Change* 39 (3), 607-50.

Chambers, R. (1992). The Self-Deceiving State, *IDS Bulletin* 23(4): 31-41.

Development Communication Report (1978). *An Interview with Daniel Benor*, No. 22, April, pp. 8-10. World Bank, Washington, D.C.

Feder, G. and R. Slade. 1986a. "The Impact of Agricultural Extension: The Training and Visit System in India", *World Bank Research Observer* 1(2): 139-161.

Feder, G. and R. Slade. 1986b. "A Comparative Analysis of Some Aspects of the Training and Visit System of Agricultural Extension in India", *Journal of Development Studies* 22(2): 409-428.

Feder, G., L.J. Lau, and R. Slade. 1987. "Does Agricultural Extension Pay? The Training and Visit System in Northwest India", *American Journal of Agricultural Economics* 69 (3): 688-686.

Feder, G., R. Slade, and A. Sundaram. 1986. "The Training and Visit Extension System: An Analysis of Operations and Effects", *Agricultural Administration* 21 (1): 33-59.

Feder, G., A. Willett, and W. Zijp. 2001. "Agricultural Extension: Generic Challenges and the Ingredients for Solutions." In S. Wolf and D. Zilberman, eds., *Knowledge Generation and Technical Change: Institutional Innovation in Agriculture.* Boston, Mass, Kluwer.

Gautam, M. 2000. *Agricultural Extension: The Kenya Experience: An Impact Evaluation.* Washington, D.C., World Bank. Operations Evaluation Department.

Gemo, H., C.K. Eicher, and S. Teclemariam, 2005. Mozambique's Experience in Building a National Extension System. Michigan State University, East Lansing.

Howell, J. 1982a. *Managing Agricultural Extension: the T and V Systems in Practice*, Discussion Paper No. 8, Agricultural Administration Unit, Overseas Development Institute, London, the Institute.

Howell, J. 1982b. 'Responses to Discussion Paper No. 8, "Managing Agricultural Extension: T and V System in Practice"', *Agricultural*

Administration Network Newsletter No. 9, London, Overseas Development Institute.

Howell, J. 1983. *Strategy and Practice in the T and V System of Agricultural Extension*, Discussion paper No. 10, Agricultural Administration Network, Overseas Development Institute, London, the Institute.

Hussain, S.S., D. Byerlee, and P.W. Heisey. 1994. "Impacts of the Training and Visit Extension System on Farmers' knowledge and Adoption of Technology: Evidence from Pakistan", *Agricultural Economics* 10(1): 39-47.

Lipton, M. with R. Longhurst. 1989. *New Seeds and Poor People*. Johns Hopkins University Press, Baltimore.

Martens, B., U. Mumert, P. Murrell, and P. Seabright. 2002. *The Institutional Economics of Foreign Aid.* Cambridge University Press.

Moore, M. 1984. "Institutional Development, the World Bank, and India's New Agricultural Extension Program", *Journal of Development Studies* 20(4): 303-17.

Mureithi, J. and J.R. Anderson. 2004. Farmer-Extension-Research Interfaces, In C. Ndiritu, J. Lynam, A. Mbabu, eds., *Transformation of Agricultural Research Systems in Africa,* East Lansing, Michigan State University Press.

Operations Evaluation Department (OED), the World Bank, 1984. *Agricultural Research and Extension: An Evaluation of the World Bank Experience,* Washington, D.C.

Picciotto, R. and J.R. Anderson. 1997. "Reconsidering Agricultural Extension", *World Bank Research Observer* 12(2): 249-59.

Purcell, D.L. and J.R. Anderson. 1997. *Agricultural Extension and Research: Achievements and Problems in National Systems.* Washington, D.C.: World Bank.

Rivera, W.M., M. Kalim Qamar, and L. Van Crowder. 2001. *Agricultural and Rural Extension Worldwide: Options for Institutional Reform in Developing Countries*. Rome: Food and Agricultural Organization.

Roberts, N. ed. (1989). Agricultural Extension in Africa. A World Bank Symposium, World Bank, Washington D.C.

Slade, R., G. Feder, and R. Chhikara. 1988, "Reforming Agricultural Extension: The Training and Visit System in India", *Quarterly Journal of International Agriculture* 27(3/4): 228-246.

World Bank. 2000. *Agricultural Extension: The Next Step*. Agricultural and Rural Development Department, World Bank Policy and Research Series no. 13. Washington, D.C.

Part 2

Commercialization of farming and privatization of extension

7

Commercialization of Agriculture and the Role of Agricultural Extension Services in Asian Nations

Mohammed Hassanullah

Introduction

Asian agriculture is now at the critical juncture of transformation from subsistence to commercial agriculture. Farmers are now increasingly producing more to sell in the market than to consume at home. With the increasing processing and export facilities contract production of desired quality of raw materials has been taking roots in almost all countries of the region. Some countries are progressing faster than the others. At this stage the public, private and NGO sector extension service providers have to play the most critical role of helping farmers as well as processors, traders and all other who are directly or indirectly associated with value chain become competitive in the global market. Global competitiveness is essential even in the domestic market in the present context of free market economy. They need to increase productivity and quality of products at the same time reduce costs parallel to their competitors. The farmers will have to think globally and act locally to safeguard their business interest rather than merely meeting the family needs. Are the traditional agricultural extension services prepared to perform these new roles?

Any system becomes irrelevant, degenerated or even dissolved if it does not change its responses as the context changes. Extension systems in Asian countries remains to a great extent non-responsive to the demand of its clients and thus it has become static and often irrelevant. Therefore, it is in the crossroad of its existence in many countries. Both the donors as well as the governments have been trying to find alternatives to the traditional public sector agriculture extension system. So far none of the alternatives, however, proved to be entirely satisfactory to meet the requirement of an dynamic system which changes with the changing context. This chapter of the book explains the nature and demand of commercial agriculture, the limitations of the traditional agricultural extension services and the means and ways to re-organize the systems and procedures so as to meet the demand of commercial agriculture.

Subsistence Agriculture

Subsistence Commercial Continuum

In the Asian countries the agriculture itself is changing from subsistence to a full blown commercial enterprise. In fact subsistence-commercialization is a continuum. When all activities of production, processing, packaging and storage are performed 100%

for home consumption it is a pure subsistence agriculture and when 100% is for sale it is a fully blown commercial agriculture. Ever since the emergence of agriculture it is moving toward commercialization because non-agricultural population is increasing and their demand for fresh and processed agricultural commodities is also increasing. There is a dilemma among the professionals, administrators and politicians as to whether one should remain or support to remain subsistence or opt and support for commercialization and endeavor to quickly transform the subsistence to commercial agriculture. In spite of this dilemma the processes of transformation is continuing either slowly or rapidly. Some countries or communities are fully commercialized, some are moving fast and some are staggering due to strong and persistent dilemma. In fact sooner the transformation process complete the better for realizing the agricultural growth potential of a family, a community or a nation.

It is a global process. But it is not so free as it is claimed or perceived to be. Resistance is enormous. At all levels of hierarchy a group of people discourage or remain aloof to promote or support or often aggressively resists transformation in the name of preserving taste, preference, tradition, culture, heritage or hitherto environment and bio-diversity. Transaction and movement of commodities pass through many difficulties ranging from local taboo to formal and informal rules and procedures of many agencies, users or the state policies and legislation adopted to protect the so-called national or community interest. In spite of those enormous resistances there is irresistible initiatives at local, national and global levels to help achieve the transformation of subsistence to commercial agriculture.

Nature of Commercial Agriculture

Agriculture has been turning to be more commercialized, i. e. people will produce to sell, not to merely eat at home. It has already gained popularity in poultry, livestock and fisheries. It is emerging for fruits, vegetables, seeds and some of field crops like soybean, maize, sunflower, cassava, sorghum, etc. People have begun to produce varieties or types of crops which are specially suitable for industrial processing and local trading as well as for export markets through contractual arrangements. This phenomenon will be more prevalent in the immediate future.

The farms will be more specialized i.e. producing a single or very few commodity(ies) to maximize the benefits of the comparative advantage of the farm or will integrate only a few commodities to maximize the use of different resources and equipments including by-products and wastes of the farm. Farmers of specialized farms will acquire a very high level of technical capability and operational experiences. As a result, they will expect to get advice on advanced technologies and management practices.

Commercial and specialized farms will be capital intensive as opposed to labour intensive subsistence farming, to maximize profitability of the enterprises. Labor is no longer cheaper in the peak operational period. In other words, farms will use more and more machines for tilling, planting, weeding, harvesting and threshing.

Demand for Services

Information of Market and Technology: Commercial, specialized and capital intensive farming will demand the most accurate information of market and technologies to avoid risk of failure and to ensure supply of the produce. The farms which are emerging now try to establish direct contacts with the sources of information as they can no longer rely on the so called extension agents. The demand for such information of markets and technologies is increasing very fast.

Quick and Technically Sound Advices: Commercial specialized and capital intensive farming involves a very high level of investment. As a result, it demands very quick solutions to production and marketing problems. It, therefore, requires quick and technically sound advice from reliable sources either of public or private agencies. It reduces risk and maximizes production. Such farmers will not hesitate to make reasonable payments for the services, if it really benefits. The necessary investments make it difficult for poor families to become commercial farmers.

Analytical Services: As farming moves toward high investments, farmers can not remain satisfied with idea and perception based extension advices unless such advices are based on analytical findings. They would not like to invest money unless they are convinced that something is really essential and likely to cause substantial loss or profit. They would like to see the scientific basis of the expert prescriptions. Farmers are now demanding soil analysis, plant analysis, analysis of available fertilizers, dead poultry birds for disease identification, etc. This demand will increase tremendously in the coming years. The extension system must be prepared and adequately equipped to meet these demands for services at reasonable costs with accuracy, reliability, and timeliness.

New Paradigm of Agricultural Extension

Importance of Agricultural Extension Work

Agricultural extension work is becoming increasingly more important in the Asian countries because demands for technological education and guidance to solve technical and marketing problems are increasing very rapidly as more and more educated people are taking agricultural professions as business undertakings. People are now investing more money in agriculture for commercial purposes. At the same time the gap between the potential and actual production, income and standards of living is also rapidly increasing. This gap must be reduced to make agriculture more and more competitive to face overwhelming influence of global free market economy. The situation is now more demanding to make the agricultural extension services more effective to assist people make the right choices of products, technologies and markets.

Client Base of Agriculture Extension Services

As the agriculture moves to be commercial the client base of agricultural extension services increases because a broad based clientele is involved in the entire value chain of commercial commodities. As a result a wide range of principles, methods and technologies of diversified clientele and diversified products becomes essential to satisfy the needs and priorities. Focus is more towards means and ways of handling second generation problems of preservation, storage, packaging, transportation and financial transaction. The system will focus on technologies that would increase productivity, reduce cost at the same time meet quality requirement as demanded by the markets. A delicate balance has to be maintained. Besides the client will demand highly reliable advices based on factual and analytical review customized to the need of an individual enterprise.

Role of Agricultural Extension in Emerging Situation

The important question is what is the role of agricultural extension in this process of transformation. Before answering this crucial question/issue there is a need to comprehend the role of agricultural extension emerged in the context of a national urge to assist farmers learn and apply the latest science and technologies to increase productivity to sustain their livelihoods in rural areas when their fellow citizen working in trade and industries were rapidly progressing taking advantage of science and technologies leading toward higher productivity and income and live better lives. The successful application of the concept of agricultural extension brought revolutionary changes in agricultural productivity and income in developed countries. Farmers increased production many fold which requires increased demand in the markets. In order to reach the consumers some of the primary commodities needs to be processed and preserved for year round sales. In these secondary arena of activities a wide range of activities emerged broadening the client base from farmers to traders, packagers, stockiest, processors and venders. Along with the expansion of client base technological base has also changed from mere production technologies to technologies of preservation, processing, packaging and transportation. In this process the generic role of agricultural extension did not change. It continues to educate farmers and the farming related people on knowledge and skills required to social, technical and economic changes in the production and consumption system. There is change of technologies as well as methods of teaching from merely personal to electronic communication media.

The extension system which could broaden its client and technological changes affecting the operation of the enterprises linked in the value chain of agriculture remains relevant to its mandate as well as demands of its clients and could sustain and expand its operation. Those who could not respond to these changes are loosing their importance and support as well as their demand is increasingly dwindling down to the bottom. The governments, sponsors and donors increasingly identify and support the alternative extension service providers viz. NGOs, business houses, industries and the community organizations during last one and a half decade. However none has come out with distinctly recognizable success because each has serious limitation in providing services. First the traditional government organizations due to increasing bureaucratization characterized by standardized operational system and procedures, rewards and punishment, orientation and attitude of public sector agencies are simply becoming increasingly more and more non-responsive to the changing needs of their clients despite increasing multi-donor support to agriculture extension. It is a colossal failure of social, political and professional leaders to keep public services effective due to lack of good governance. The issue is very broad relating to the entire public sector services of a country. Secondly, the NGOs, the donors are pumping large volume of resources through a large number of international, national and local NGOs, who are more attentive to accumulation of resources than to employ and professional staff for providing services of desired quality and standards. Contrary they are rather transforming themselves as highly bureaucratic, non-transparent and authoritarian just opposite to what they preach in order to get access to fund during last three decades. The third alternative the business houses and the industries are supposed to provide embedded services along with their sales of inputs or contract for supply of raw materials either for processing or export. From early nineties it is promoted and supported by the donor supported development projects through so called Business Development Service (BDS) providers as well as to some extent by the government also. Recent reviews show that these initiatives resulted an emergence of a *dadan*[1] and bonded labor system. The *dadan* and bonded wage providers enjoy a high share of profit earned by capital but do not bear responsibility of risk of failure of the business either for poor inputs provided, management failure or natural disasters. The system is exploiting both Small and Medium Enterprises (SMEs) and labors. The so-called corporate social responsibilities did not take roots in the society. We are heading toward a feudal business system of depriving both producers and labors and protecting the unethical accumulation of wealth/resources which is protected through private sector security service providers established by retired public sector police or army officials. In this circumstance there is a need to examine the alternative system of providing extension services.

---

1. It is a system of advance purchase of produces at low prices during a socio-economic distress or sales of inputs, often low quality of inputs on credit at high prices when unable to buy on cash. The *dadan* providers siphon off the major benefits without sharing the risk of failure or poor performance due to supply of poor quality inputs, wrong advices or poor management or natural disasters. Failure for any reason make the operator perpetually dependent on the *dadan* providers. They take advantage when institutional credit system become ineffective and non-responsive to credit need of the commercial operators.

Asian countries has a long tradition of community support system for transfer of traditional technologies to its successive generations as well as to support members of the community to sustain livelihoods. The colonial regimes created institutional base in late nineteenth and early twentieth centuries to provide European style agricultural extension services in different sub-sectors of agriculture independently. The concept of out of school education of working people in agriculture was a super imposition during post world war II when technical assistance program of USA intensified in almost all Asian countries. The major focus was to help farmers sustain livelihoods through increasing productivity of subsistence agriculture. Never there was an attempt of exploiting production potential of the available resources.

Need for a Basic Change

The ineffectiveness of past efforts may be broadly attributed to failure of; (i) assuming appropriate roles of extension agents, (ii) having an appropriate structure of agriculture extension services, (iii) adopting an appropriate policy framework for agricultural development, and (iv) transferring ownership to the clients through their organizations. Firstly, it is a mistake that extension was perceived as an information distribution system in most of the third world countries. As a result, many messengers, not educators, were recruited and mobilized to pass on technological information to innovators, contact farmers and group leaders either through individual or through the group contacts. Secondly, it is taken as guaranteed that irrespective of any kind of structure of the organizations, extension workers can work effectively if resources are pumped in. Thirdly, it is also not appropriate to think that irrespective of any economic and fiscal policy, extension can influence farmers to bring desirable changes in agriculture. Fourthly, no extension system is expected to become transparent and accountable to its clients in order to meet their needs and aspirations unless the ownership of the system remains with clients through their own organizations, whether it is public, private or NGO. As a result, extension has not been as effective as desired in spite of massive staff increase, finance, training and logistic supports financed mostly through grants and loans of national and international agencies during last four decades. In fact, the organizations either public, private or NGOs have expanded their systems with traditional civil service orientations and structures with bureaucratic values and attitudes. It is now time that we re-examine the systems and redesign them in such a way that becomes meaningful, responsive, effective and affordable to the farmers and related population in meeting the challenges of the commercial agriculture. Mere reduction of staff or withdrawal of operation in favor of different NGOs as often prescribed by funding agencies is not likely to increase effectiveness of extension work in any way.

Challenges of Extension Services

Capable Extension Personnel

As demands for technical services with desired level of accuracy, reliability and timeliness at reasonable costs increases, the employment of really capable extension workers, in particular, and extension personnel including administrators and specialists, in general, becomes very much essential. The commercial farmers and agricultural enterprises will in future demand even higher level of knowledge and skills. Extension services in the commercial agriculture must gain a very high level of practical knowledge and skills to satisfy demands of fixing anything in the field, farm or industry. It may be mentioned that USA started extension work with a degree of BS in Agriculture and later made it compulsory to do MS and Ph Ds, having the status of professors. Some developing countries are now changing the technical field workers to graduates of agriculture. It is not unlikely that a few technicians may gain high degree of skills but that does not improve the entire system. The situation is worse in the so-called NGO extension service provider. They employ often high school graduates of science or art undergraduates for disbursement and realization of micro credit. Experts often ironically explain that technical advices are embedded in it. Funding agencies often provide them profuse number of very short trainings to make them technically capable. It has been observed that they often remain 50% time absent in the working station to attend short trainings offered by many different agencies on many different issues. There is no research evidence that basic skills can be developed by a large number of short trainings. Many such so-called trainings are found to be a few hour discussion meeting. A more economic option is to rather employ appropriately educated persons. Graduates have the potential to grow professionally and provide services of worth. Commercial operators are not going to pay for the services unless such services can provide a lucrative dividend in terms of increasing productivity or income. Since role of extension agents is going to change from information distribution under subsistence farming to advice and education for solving problems of commercial agriculture the change needs to be faster to make the farmer competitive in the global market economy.

Training

Changes in the roles of agricultural extension requires training and retraining of extension staff to maintain them as truly knowledgeable with and skilful in the application of the new technologies of bringing technical and socio-economic changes in the client system. But one should be careful as to how much time an employee should spend for his/her own professional development. Ideally one should not spend more than 10% time for self development. Careful planning is necessary as to what knowledge and skills can be transmitted through communication system, what should be communicated through discussion and dialogue and what needs to be transmitted through training. It should be noted that capability of persons, lacking basic knowledge and skills, can not be improved much by frequent training. Frequent withdrawal of staff from work situation rather destabilize work ethics and devotion to work. However, short, in-service training needs to be organized for orienting new recruits, when there is change of program perspectives, new technological development, etc. A balanced program needs to be pursued to provide equitable opportunity of learning new developments.

Reliable Technical Advices

Another important challenge of the commercial agriculture extension is to make the technical advice highly reliable. Traditionally, thousands of so-called extension agents were employed with no office, no logistics, no vehicle even sometimes not a by-cycle, no equipment, no facilities for testing and measurement of anything. They are supposed to roam around the villages and advice the farmers on production of enormously different commodities and products and their consumption. They may be termed as *Scavenging Extension Agent*. In most cases they spread their ideas, impact points as in case of T&V system or the information as communicated by their superiors. It is not unlikely that such advices might often caused failure causing economic losses to the farmers. During the last half a century enormous resources were poured in to improve the system. Much of those resources were absorbed by higher echelon of the organizations at international, national, regional and local administrative levels. Operations at the grass roots level, the extension agents, however, did not change much either in public, private and NGOs. All those are shockingly absorbed by subsistence farmers. After all it would not give much benefit than mere earning subsistence living. In commercial agriculture people would like to see that advice really works as promised because it is capital intensive. Failure would cause substantial losses. Without reliability of advices, no one would like to make an investment in commercial agriculture or pay for extension services.

Commercially Viable Educational Activities

All educational activities such as meetings, demonstrations, field days, trainings, etc. need to be of very high level of usefulness, so that anybody missing participation would be remorseful of being deprived of knowledge and enlightment. Many of them are willing to pay for participation in such activities. It is a challenge of the extension service as how to make the activities really useful and hence commercially viable. Only then participation becomes worthwhile and clients would be willing to pay for the services.

Business Orientation of Extension Personnel

Subsistence farming may sustain living on farm but has no scope to provide means to earn more, invest more and live a better life. Farmers and people associated with farming or providing services to farmers need to develop the ability to conduct a business. In order to help them acquire knowledge and skills for running the business, the extension workers themselves need to be oriented and knowledgeable about farming business. In commercial agriculture, it would be a great challenge on how to inculcate business orientation among the extension personnel as they are so long accustomed with charity and social welfare activities. Their college or university education does not include business courses too.

Principles of Organizing an Effective System

In order to make the agricultural extension services able to meet the demands of commercial agriculture, the system needs to be organized on the basis of certain principles which have empirical validation or at least some logical probability of success. Ten general principles are enunciated here as guidelines.

Principle I: Expand scope of work to affect all aspects of life and profession of the farming community

Problems of clients know no bound. The extension agents should, therefore, work with anything that stands in the way of productivity, income and living of their clients. Within a given operational area, an extension agent should be given the mandate to work for agriculture production, marketing, farm management, conservation, development and use of resources, environment upgradation, disaster management, home and youth development and institution and leadership development. Agriculture should be considered in its broadest sense of all activities envisaged in a "Farm Production Plan" of a farm family. Given this broad mandate. the extension agents should develop a plan of work by sub-sectors of agriculture, technology and the client segments for a given period of time. Only then extension agents can develop confidence and credibility among their clients and can make significant influence in bringing desirable changes in the farming community.

Principle II: Adopt problem solving approach as opposed to communication or technology diffusion approach

There are five distinctly distinguishable approaches of agricultural extension namely, social marketing approach, technology diffusion (transfer) approach, communication approach, teaching learning approach and problem solving approach. Each of these approaches has a distinctly different way of accomplishing the extension function (Hassanullah, 1995, pp. 40-53). Among these approaches, problem solving is the most appealing approach to the clients because it emerges from a felt need and proceeds with active participation of clients at all stages of planning and execution of a programs. It provides a more flexible and broad based operational framework. Since it involves people, it necessitates establishment of formal or informal functional groups and extension agents assume a helping role that informs people about problems and their solutions, advises them to make the right choices, motivates and organizes them for action, demonstrates what to do, coordinates activities and leads them to move toward the solution of the problem. People prefer to solve their problems more than to merely be informed, learned, adopt or buy a technology with somewhat vague or diffused perception of their purposes or benefits which may or may not relate to their problems. Since most extension services of the Asian countries are controlled by the Ministries of Agriculture, they are designed to implement the policies and programs of the government. In order to be effective it must reverse its basis of work from a wish of the government to the problems of the clients.

Principle III: Expand clientele system to people associated with a problem and make segments of the clients for program administration

The clientele of an extension agent covers a wide range of persons who are directly or indirectly associated with a particular problem. They are enormously heterogeneous. In order to solve a problem, the needs and aspirations of different segments of clients should be taken into consideration though they are widely different. It is also socio-economically more desirable to give certain segments more importance than the others. There are also wide variations among technologies, and not all technologies are relevant to all segments of the clients. To utilize the time and resources or the extension agents properly clients need to be segmented so that separate programs can be organized for different segments of the clients to solve a particular problem.

Principle IV: Assume full educational/advisory role of an extension agent

In recent years, the role of an extension agent has been grossly changed. In the T & V system an extension agent instructs farmers what to do and what not to do at all times of the year as if he/she is the Manager while the farmers are workers. In other systems extension agents, are designated as the communicators. It is rare in developing countries to think of Extension Agents as educators who inform, advise, motivate, organize, coordinate, demonstrate and lead people toward desirable changes to solve a problem or any aspect of the life of farmers and the farming community. In order to be effective in performing these roles, he/she has to plan, conduct, evaluate, record and report about the roles to be performed as an educator/advisor. Graphically, the roles of an extension agents should be as follows:

Educational Roles	**Major Sub-roles**				
	Plan	**Conduct**	**Evaluate**	**Record**	**Report**
Educator					
Informer					
Advisor					
Motivator					
Organizer					
Coordinator					
Demonstrator					
Leader					

To be effective, the educational roles of the agricultural extension agents must be established within the operational framework of problem solving approach.

Principle V: Develop national agricultural extension service as a locally or regionally integrated system of teaching, experimentation and out-of-school education.

A locally or regionally integrated system of teaching, extension and experimentation on the life of a farming community can mutually strengthen and help clientele exploit their full potential of higher productivity, income and level of living within the limits of their agro-ecological potential. The application of agricultural science has wide variations with location specificity. It should be locally or regionally learned and applied. Regional Bureaus, either independent or controlled by regional universities can better serve this purpose than a national agricultural extension service under the Ministries of Agriculture.

Principle VI: Appoint extension agents with high level of formal technical education and adequate social science background

The qualification of an extension agent is one of the most critical factors which makes extension system effective. Extension agents need to have capability, freedom of action and support. Of these qualifications capability is the primary requirement. Capability constitutes three component; background, educational qualification and behavior. While background and behavior are screened through selection and professional educational process, the choice of level of education of extension agents is a crucial decision of an extension system. The USA began its extension work in 1914 with people having a degree of BS in Agriculture and later made the Masters Degree as requirement to get tenure service in extension. Japan changed its technician extension workers with Bachelors of Agriculture as early as late fifties. Some developing countries employ graduates as extension agents. In most developing countries the number of extension workers with high school qualification has increased manifold, but no attention was given to employ higher qualified persons. Those poorly qualified staff were offered many short-term training but in vain. Asian countries now should start the process of changing technician extension agents to graduate extension agents and appoint them as Farm Advisors, not as Supervisors or Communicators or Demonstrators. Without such education, the Farm Advisors will not be able to assume the educator/advisor role and meet the demands of capital intensive, commercial and specialized agriculture of the this century.

Principle VII: Adopt a task structure based on the extension education model

What should an extension agent do in a problem solving situation? Often we prescribe a model and determine what an extension agent should do either weekly, fortnightly or monthly and often prepare a full calendar of work for a year. A simple and immitigable pattern of activities is a model. The most commonly prescribed and distinguishable models are service (input) delivery, advisory service, model farmer, training & visit, functional literacy, participatory group models (Hassanullah, 1995, pp 58-65). Each has a distinctly different pattern of activities. In a problem solving situation it is not feasible to comprehend what an extension agent would do for any period of time. She/he will have to follow the extension education model which depicts that an extension agent will perform all educational activities required to solve a problem along with farmers using the principle of "as and when required". She/he will have to identify the problem(s) and prepare a plan of work to solve the identified problem(s), spell out a calendar of activities which may range from a personal visit to a conference or mass media programs in consultation with the clients. In this exercise she/he is assisted by his/her supervisor or the subject matter specialists. Once such calendar is prepared, it becomes the basis of supervising their work.

Extension provides common formats to write the plan of work in a way that helps keep account of the time and resources spent for different groups of clients, technologies, sectors and sub-sector as required to satisfy the policies of the sponsoring agencies.

Principle VIII: Adopt a completely decentralized and reduced hierarchical structure of the organization

A centralized organization is by no means suitable for administering problem based educational programs at grass roots level and meeting the demand of commercial agriculture. It requires a completely decentralized operation with reduced hierarchy to make it quickly responsive to the demands of the commercial operators.

Principle IX: Adopt a system of completely grassroots planning and implementation of programs both for commercial and social welfare objectives

Centrally determined and enforced pattern of activities have proved to be ineffective in agricultural extension. In order to meet the demand of commercial operators as well as the welfare need of a locality, the extension program should be prepared and implemented by the Farm Advisors at the very Extension Working Unit level. It however can be aggregated at different levels of operation for administrative purpose.

Principle X: Ensure client ownership of the system and their participation in program management

Whoever finances it the system must be owned by the clients through their own organization in order to establish accountability and transparency. Client participation is not merely receiving the services but also in making program decisions and implementing them. This ensures that the plan of work of extension agents meets the needs and aspirations of the clients. There should be an institutional arrangement in the local governments or farmer organizations that allows the exercise of power and authority to control operation of the extension programs and to meet the needs and aspiration of the clients of the extension system. To make these principles function at tandem to strengthen and smoothen the agricultural extension service, necessary changes may be required in organizational structure, staffing, supports and services (Hassanullah, 995).

Governance

Finance

At present in many countries over 90% of the extension budget is spent for salaries and very little is left for financing educational programs. Administrative cost of any program should not exceed 20% and 80% should be spent for operational costs of the educational programs. It is because of poor and irrational financing that the magnitude of educational activities is very low often insignificant to affect any substantive change in the client system. An extension agent should implement an educational program costing about five times of the cost of maintaining herself/himself.

Besides the government funds the bureau should have the authority to accept public donations, subscriptions and fees for its services. It should create and operate trust funds for financing special extension programs to meet the special interest groups or to meet social objectives.

Administration

Agricultural activities are time and season bound. In order to be effective, extension service activities should also be matched with agricultural activities. Directions, sanctions, approvals and appraisals must, therefore, be clear, decisive and in time. Much of our operational activities fail because of delayed actions, inaction or confused administrative actions to develop and carry out programs at grassroots level.

Supervision and Guidance

Much of our extension supervision is post mortem examination of success or failures. Often proper supervision and guidance is a great support to keep extension agent effective. Supervisors demands to organize some activities to present themselves in front of the clients. In such occasion assembling of people is considered as credibility of the extension workers. It does not help much to increase capability of extension workers and hence the performance of extension system. It should be participatory i.e. the supervisors should be attending the educational activities and guide the agents improve the conduct of such activities effectively.

Structural Adjustments of the Existing Systems

Marginal improvements in funding of on-going activities, addition or reduction of few staff, some logistic, some training are not expected to improve the extension system to meet the needs and challenges of the commercial agriculture. The agricultural extension system needs radical changes in structure, functions and conduct. Some important and basic changes are briefly discussed.

Privatization of Potential Commercial Activities

Historically all field organizations of agriculture now have provision for some kinds of production, inputs, credits administration, extension, quality control and governance functions. Activities of NGOs are more diffused. As a result, none of the activities is effectively performed. All potential commercial activities of all field organizations of all sub-sectors of agriculture need to be privatized except which are needed to be used for demonstration purpose. This will help reduce waste of fund and staff time, and sharply focus on the development of skills for a particular type of function. The size of enterprises, kept for demonstration, should be such as to prove its commercial viability and should be managed in such a way as to make a real demonstration and use it very frequently for educational purposes.

Integration of Control Functions

Control and governance functions of agriculture need to be consolidated in one organization with a new mandate and staffing pattern so that it develops skills and facilities and have full authority to exercise laws governing agricultural activities and enforce quality control of both inputs and outputs. Extension officials should not be given a control function, because by exercising this function properly they may loose the trust of their farmers and farmers may not be willing to talk freely about their problems.

Local Integration of Extension Like Functions

Upon privatization of commercial functions and integration of control functions, all extension and extension like functions of research and development organizations of all sub-sectors of agriculture need to be locally integrated in an extension system to be governed by local bodies representing the different sections of the client system under the direct administration of the local bodies formed by legislation or of educational institutions where feasible. Regional bureaus are the appropriate structure to be locally organized and governed by local governments keeping opportunities for career development.

Planning and Implementing an Annual Plan of Work

Planning must be organized at grass roots level with participation of the representatives of the client system. In other words, the Farm Advisors with the assistance of Agricultural and Rural Extension Committees of the local governments will have to plan the program following a standard format of writing the plan. Such format must account for the time and resources spent for different client segments, program areas and activities performed. It must have statements of problems identified, technologies chosen, rate of diffusion aimed at and activities to be performed in a time. Once plan of work of each Farm Advisor is prepared it should be aggregated at the supervisory level and any educational activities to be performed by the supervisory staff be added and the extension plan of work is prepared which should be approved by Agricultural and Rural Extension Committees of the local governments/Councils. In the same way the district and regional extension plans of work could be prepared by compiling the approved plan of work. It can also be compiled at national level by the Ministry of Agriculture or by the National Council of Agricultural and Rural Extension as a matter of evaluation and monitoring purpose.

Each Farm Advisor and other extension staff at different levels should schedule their planned activities and perform with the help of the members of the committees, if and when any public support is needed. Implementation of the plan should then be monitored periodically by collecting information through a standard formats.

Associated Changes

Changing Academic Curriculum of B Sc (Ag) Programme

The present curriculum of higher agricultural education of some Asian countries was designed on the concept that extension work will be performed by the persons having diploma in agriculture consisting of crops, livestock, forest and fisheries under the control of the Department of Agricultural Extension and the line organizations of other sub-sectors of agriculture will provide specialist services to them. As a result, specialized B Sc (Ag) curricula were introduced. None of the graduates of these specialized curricula is adequately prepared to be employed as Farm Advisor. A general curriculum of agriculture combining all relevant disciplines with a reasonable ratio are to be developed and administered to produce B S in Agriculture who will be suitable for employment as Farm Advisors. Persons having masters in different disciplines can then work as specialists of the extension system.

Introduction of In-service B Sc (Ag) Programme

Some of the capable and experienced Block Supervisors of some countries fulfilling the basic science requirements should be given opportunity to prepare themselves as capable Farm Advisors through an in-service BS (Ag) curriculum. If necessary some intensive basic science courses can also be provided to meet the requirements. Some of the ATIs should be converted as colleges of agriculture under certain regional agricultural universities to administer these programs.

Upgrading Colleges into Faculties or Universities

Agricultural universities provide the best institutional framework to integrate extension functions at regional level and have operational flexibility to meet the needs of different sections of client system to attain both commercial and social objectives of agricultural extension. Different agricultural, veterinary and forest colleges should be upgraded as agricultural universities with changed mandate of teaching, research and extension work with legislative support. Without this integration none of the institutional framework is expected to be effective.

Conclusion

Agricultural extension services has become increasingly important in the Asian countries as their subsistence agriculture is gradually transforming to commercial agriculture. Commercial agriculture is capital intensive and to some extent specialized operated more and more by educated people. It demands quick and technically sound advice, based on scientific analysis in order to remain capable to produce products which are globally competitive in terms of productivity, quality and costs. It demands extension services which are reliable and pay a lucrative dividend when applied and are available at call. The existing agriculture extension system is not capable to meet these demands. The major challenges of the system are; i) how to increase capability of extension personnel, ii) how to make them available to clients at call, iii) how to make their technical advice reliable, iv) how to make their education activities commercially worthy and commercially viable and v) how to inculcate business orientation among the extension personnel. The agricultural extension services can meet these demands, if organized, on the basis of ten principles as suggested in the present chapter with necessary structural and functional changes and reorientation.

References

CPD (1995). Experiences With Economic Reform: a Review of Bangladesh's Development 1995, Dhaka, Centre for Policy Dialogue/University Press Ltd.

Hassanullah, M. (1995). Managing Extension Services. Dhaka, University Press Ltd., pp. 40-53).

Hassanullah, M., (1996). Muzaffer Ahmad, Hafiz G. A. Siddiqui and Shamsul Haq. 1996. Performance of agricultural Extension Organization of Bangladesh. Dhaka, University Press Ltd., pp-48-49.

Hassanullah, M. (1989). An Analysis of Out of School Educational Process of Agricultural Extension Services of Bangladesh. *Bangladesh Journal of Extension Education. Vol. 4 No. 1 & 2, Jan-Dec.*

Mahatab, F. et.al., (1992). *Report on the Evaluation Team for Extension and Research Project II.* Dhaka, Ministry of Agriculture. p 110).

Phillips, S. et al., (1991). *Post Evaluation of Extension and Research Project II. Dhaka,* Food and Agricultural Organization of the United Nations. P. 37.

Swanson, B. E. (1990). *Report of the Global Consultation on Agricultural Extension.* Rome, Food and Agricultural Organization of the United Nations.

8

Developing a Decentralized, Market-Driven Extension System in India: The ATMA Model

J. P. Singh, Burton E. Swanson, K. M. Singh

Introduction

India's agricultural extension system is at a pivotal point in its evolution. Since independence, the extension system has focused on four major strategies, reflecting the dominant agricultural and rural development goals during each period.[1] In looking back, the evidence suggests that investments in agricultural research and extension have served the country well, particularly in achieving food self-sufficiency. At the same time, hunger and malnutrition are persistent problems for the rural poor, and rural economic growth seems stalled at the 2% level as the rest of the economy moves forward at a quicker pace. This is the backdrop in considering the future direction for agricultural extension in India.

The 100 million farm families in India are supported by a large agricultural extension service financed by

State governments. This service has had over 100,000 staff members. Since independence it has used different extension approaches with mixed results (Samanta & Hegde, 1998; Swanson & Mathur, 2003). Extension in India is the responsibility of the State government with some support from the Central government. The next level is the District, which has on the average 200 000 farm families and consists on the average of 25-30 blocks. Each block has about 40-50 villages.

The introduction of the Training-and-Visit (T&V) Extension system was an important milestone in the history of extension in India. The basic premise of T&V was that there was enough technology available awaiting diffusion to and adoption by farmers. The T&V Extension system was first introduced in 1974–75 on a pilot basis in the Chambal Command area of Rajasthan and Madhya Pradesh. Based on positive feedback, the project was further extended to 17 other states in 1978–79. Thus the multi-purpose approach that started during the early 1950s under the Community Development Programme (CDP) approach to extension was gradually replaced by a single-line of command extension system that focused on the major food grains toward the national goal of food security.

Problems and Constraints

The Training and Visit (T&V) Extension System was effective in disseminating *Green Revolution* technology, especially in the high potential, irrigated areas, but it had little effect on the productivity and incomes among farmers in rainfed areas. In addition, by the early 1990s,

---

• This paper was partially prepared with funding provided through a competitive grant from the Sustainable Agriculture Systems, Knowledge and Institutions (SASKI) Thematic Group at the World Bank, Washington, D.C.

many other systemic problems were apparent:

- The introduction of T&V Extension *greatly expanded the number of village extension workers* (VEWs) in the Department of Agriculture (DOA), resulting in long-term financial problems for state governments. Since most state government funds go for salaries, most extension activities are dominated by top-down, central government programs.
- Government's continuing focus on increasing food production resulted in *extension being commodity and supply-driven*, in contrast with a focus on diversification and farm income (i.e., *being more market-driven*).
- Dissemination of Green Revolution technology substantially increased the production of food staples; therefore, commodity prices fell during the 1980–90s resulting in *declining farm income*.
- The emphasis on food security during the 1960–80s resulted in an extension system that was limited to the staple food crops and dominated by the DOA. The other line departments, including Animal Husbandry (DAH), Horticulture (DOH), Fisheries (DOF), Sericulture (DOS), etc., had a very limited extension staff, virtually no extension programs and operated separately from each other. As a result, there was no integration of programs across departments (i.e., lack of a *"farming systems" approach*)
- By the 1990s, the line departments primarily focused on the distribution of centrally funded subsidies and inputs. This situation had the following effect:
  - Line department staff became increasingly *accountable to government*, rather than to farmers;

- Since government was partially involved in input supply, the government field staff viewed private input supply dealers more as *competitors* than as partners;
- Given this focus on central government schemes, there was less need for extension to work closely with researchers, resulting in *weakening research-extension linkages*.

- Finally, with the exception of donor sponsored schemes, extension gave very little attention to *organizing farmers into groups* and, thereby, empowering farmers.

During the mid-1990s, the Government of India and the World Bank began exploring new approaches to extension that would address these system problems and constraints. The result was a new, decentralized extension approach, which would focus more directly on agricultural diversification and increasing farm income and rural employment. The central institutional innovation that emerged to address these system problems was the Agricultural Technology Management Agency or "ATMA" model that was introduced at the district level to 1) integrate extension programs across the line departments (i.e., more of a farming systems approach), 2) link research and extension activities within each district, and 3) decentralize decision-making through "bottom-up" planning procedures that would directly involve farmers and the private sector in planning and implementing extension programs at the block and district-levels. This model was pilot-tested through the Innovations for Technology Dissemination (ITD) component of a World Bank-funded, National Agricultural Technology Project (NATP) that became effective in 1998 and concluded in June 2005. The next section of this paper will describe how this decentralized extension model is organized and how it operates.

The ATMA Model

Overview. The Agricultural Technology Management Agency (ATMA) is an autonomous organization that has considerable operational flexibility. For example, it can receive and dispense government funds, enter into contracts, maintain revolving funds, collect fees and charge for services. In addition, it operates under the direction and guidance of a Governing Board (GB) that determines program priorities and assesses program impacts. The head of each ATMA, known as the Project Director or PD under the NATP, reports directly to the GB. The PD serves as chair of the ATMA Management Committee (AMC), which includes the heads of all line departments and the heads of research organizations within the district, including the Farm Science Centre, known as Krishi Vigyan Kendra (KVK) and Zonal Research Station (ZRS). Each district has a KVK; therefore, this multidisciplinary *Farm Science Center* plays a critical role in both on-farm research and training farmers in new production and value-added processing technologies. However, it is the PD that helps coordinate and integrate all agricultural research and extension activities carried out within the district. The organizational structure of the ATMA model is shown in Figure 1; the remainder of this section will explain how these different components of the ATMA model operate.

ATMA Governing Board. As mentioned above, the ATMA GB sets program priorities and provides guidance as to how research and extension programs

Figure 1: Organizational Structure of ATMA

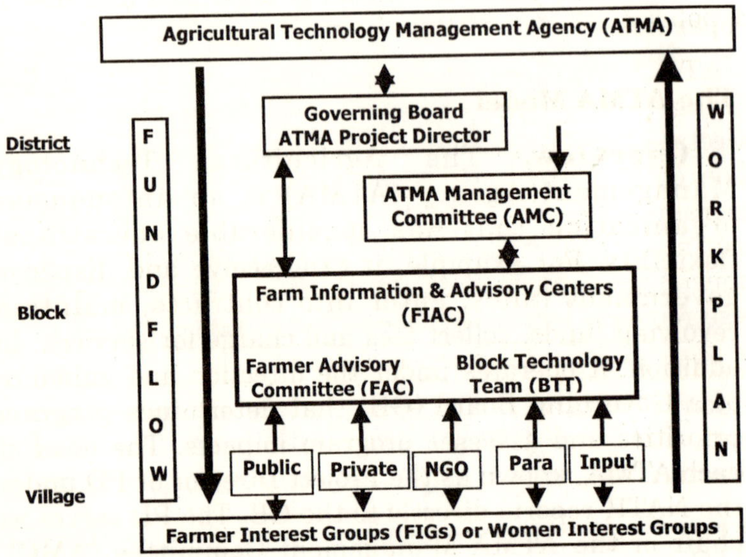

are implemented within the district. The composition of the GB provides an equal balance between the heads of the line departments and research units within the districts and the stakeholder representatives, including a cross-section of farmers, women and disadvantaged groups, and private sector firms within the district. The GB is chaired by the District Magistrate or Collector who is the highest ranking government official in the district; the ATMA PD serves as Member Secretary. The rationale for this balanced GB is to provide a platform where farmer representatives and private-sector leaders come together with agency heads to discuss and determine extension priorities. This framework for public-private dialogue provides the opportunity for both groups to learn from each other and to jointly learn by observing successes or failures of different program initiatives within the district. The primary functions of the GB are to review and approve the Strategic Research and Extension Plan (SREP) for the district (described later), to review and approve annual work plans, and to set policies and procedures for ATMA operations.

ATMA Management Committee. The ATMA Management Committee (AMC) serves as the Secretariat of the GB and helps coordinate and integrate research and extension activities within the district. Program requests come from each block and the AMC scrutinizes these requests on the basis of technical, financial and management criteria. The AMC then sends these requests to the GB for review and final approval.

ATMA Personnel. By design, the number of personnel assigned to ATMA's headquarters is very small, so this organization does not become another government agency.

Farm Information and Advisory Centers (FIACs). These centres were established at the block-level in each project district with NATP funding. These small, block-level facilities include an office for the Block Technology Team (BTT). The FIAC has become the single-window delivery mechanism for extension programs within the block. Also, the provision of Internet access serves as an important resource for all participants. For example, farmers can access up-to-date market information, while BTT members can easily communicate with headquarters and access new technical and market information.

Block Technology Teams (BTTs). In each block, a BTT was formed including one technical officer from each line department with an extension function. In general, the BTTs include technical officers from the Departments of Agriculture, Horticulture, Plant Protection, Soil Conservation, Animal Husbandry (including Veterinary Service), Fisheries, Sericulture,

Cooperatives and Marketing. The role of the BTT is to consult with the Farmer Advisory Committee (FAC) and then to develop a comprehensive extension program called a Block Action Plan (BAP) that is consistent with farmer needs. This BAP is then reviewed and approved by the FAC and submitted to the ATMA for funding. Since most extension funding flows directly from the ATMA to the BTTs, this arrangement has resulted in the BTTs being accountable to both the FAC and the ATMA GB. The net effect was to improve extension program coordination and delivery at the block level, since program responsibility and accountability has been largely delegated to the BTTs, with oversight being provided by the FACs.

Farmer Advisory Committees (FACs). A key element in this new, "bottom-up" extension planning strategy was the formation of FACs in each block. These FACs are composed entirely of farmers who represent different socio-economic categories of farmers within the block. FAC members are the heads of the FIGs. The role of the FAC is to advise the BTTs on extension priorities for the block. In addition, the FAC reviews and approves the annual BAPs prepared by the BTTs before they are submitted to the ATMA for funding. Then, the FAC monitors and provides feedback to the BTT on BAP implementation. These FACs have become an integral part of the formal feedback mechanism between farmers and the heads of the research and extension programs within the district.

Farmer Interest Groups (FIGs). One important objective of the ATMA approach was to redirect extension activities toward diversification into high-value crops and products and the overall goal of increasing farm income and rural employment. Therefore, in pursuing this market-driven approach to extension, it became essential to get farmers organized around specific crops or products where there is market demand and that are appropriate for the agro-ecological conditions and resources of different farmer groups. In addition, to successfully supply different markets, it was also essential to get these groups organized to achieve economies of scale and to create an efficient supply chain. Once these different FIGs are organized at the village level, they soon began to organize along crop or product lines as block-level farmer associations (FAs) and district-level Farm Federations (FFs).

Steps in Implementing the ATMA Model and Creating a Market-Driven Extension System

To begin implementing the ATMA concept, the first step was to engage the leaders of the research institutions and the heads of the line departments in developing a Strategic Research and Extension Plan (SREP) for the district. This process began by first training these research and extension leaders in how to conduct a participatory rural appraisal (PRA). This process joined these research and extension leaders—in many cases for the first time in a joint activity—with farmers from representative villages within the district. Collectively, they began to listen to farmer problems and then they had to translate these concerns into research and extension priorities. In the process, they examined "success stories" within the district, examples of how entrepreneurial farmers had identified markets for high-value crops/products (HVC/P) and developed supply chains to serve these markets. After completing a SWOT analysis, these research and extension leaders developed a SREP that was presented to ATMA GB for consideration and approval. This SREP helped establish research and extension priorities for the district that were then implemented through this new ATMA model.

After preparing the SREP for the district, the next step was to implement this plan using a market-driven extension approach. Based on experiences gained through NATP, it is possible to outline these steps and procedures in more detail, as outlined in Figure 2. The remainder of this section will briefly describe these steps, including examples to illustrate how these procedures were actually implemented. For a comprehensive case study that explains this process in more detail, see K.M. Singh, et al., 2005.

Figure 2: Steps in Developing Market-Driven Extension System

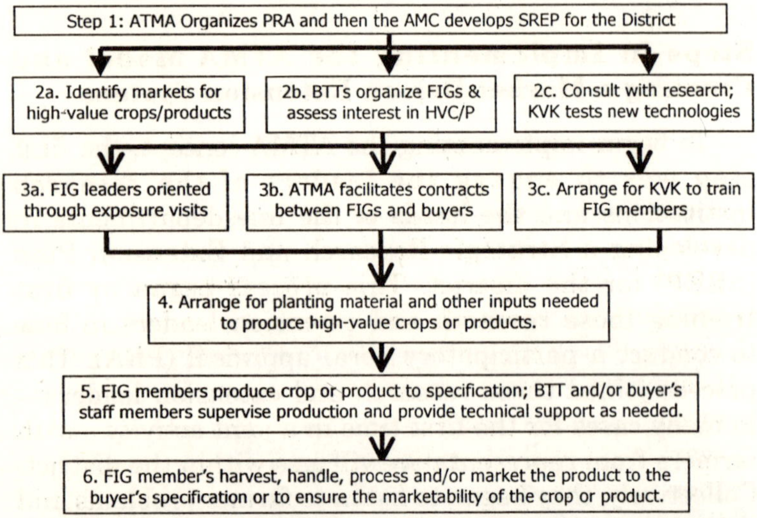

Identifying Markets and Supply Chains for Promising High-Value Crops and Products. There are four axioms that are essential to a successful "market-driven extension system."

- The *first axiom* is that if there isn't a market, don't encourage farmers to produce a specific crop or product. Therefore, the first task to be carried out is to assess the potential markets for different high-value crops or products that can be successfully produced in different blocks within the district.

- The *second axiom* is that if farmers cannot easily transport the product to market, look for more promising products that can be easily marketed.
- The *third axiom* is that if the crop (or product) cannot be successfully grown or produced within the district due to unfavorable agro-ecological conditions, then look for more promising crops or products that are well suited to the district.
- The *fourth axiom* is to diversify into a variety of different high-value crops/products that are suitable for different FIGs or WIGs within the district. This approach will mitigate risk by not saturating the market with one or two products and, thereby, driving down prices.

The first set of products, markets and supply chains to be examined are the "success stories" that are identified during the PRA. These success stories are the results of entrepreneurial farmers who have identified a market, worked out a successful production technology and a way of marketing the product. The task is to assess each success story to determine whether it is supplying a very small niche market or whether there is considerable scope to expand the production and marketing of this crop/product. If the latter is the case, then these entrepreneurial farmers can either work with the ATMA as a key resource person in training other farmers or he/she may decide to become a Farm Leader in the district who can help organize interested farmers into FIGs and then train these FIG members in the production of this crop/product and in organizing the marketing of this crop/product.

In India, with the move toward a more market-driven extension system, a portfolio of 250 different success stories was compiled from across the country under NATP (see IIM-Lucknow, 2004b). In many cases, these products can be marketed locally or to nearby regional markets, which can be replicated throughout the country. In addition, markets for higher-value products are expanding rapidly in India as the urban middle-class has more disposable income and desires new products. Therefore, markets for different high-value products can be expected to grow rapidly as India continues to experience rapid economic growth. The key for the ATMA is to first identify these different markets and then to determine if farmers within the district would have a competitive advantage in producing these crops or products due to superior growing conditions, proximity to markets and/or a suitable transportation system.

Organizing Farmer Interest Groups (FIGs). As noted earlier, different FIGs have different resources, interests and the capacity to manage or tolerate risk; therefore, different FIGs and Women's Interest Groups (WIGs) will be interested in pursuing different crops, livestock, fisheries or other enterprises (e.g., vermi-composting or mushroom production may be of interest to very poor, landless rural women). As a result, there will be a range of FIGs within each block and district, depending on their interests and resources. The key for the ATMA, working through the BTTs and FACs, is to get these different groups of farmers organized into different types of FIGs.

Consulting with Research about Available Technologies. Since researchers from the local KVK and, possibly, ZRS would be involved in the PRA and in assessing different success stories within the district, a preliminary analysis of the technical feasibility or suitability of different crop and/or livestock enterprises can be carried out while developing the SREP. However, as more detailed information is gathered on specific markets for different crops/products and farmers interests, it is important to carefully examine the agro-ecological conditions in each block of the district and the availability and suitability (through on-farm research) of different production technologies vis-ḍ-vis local growing conditions.

Orienting FIG Leaders through Exposure Visits. The NATP made extensive use of exposure visits or "farmer-to-farmer extension" in orienting different FIGs to different high-value crops, products or enterprises. If a group of FIGs were interested in a particular crop or enterprise, then the FIG leaders would be taken to visit an entrepreneurial farmer or FIG in a nearby district and/or to the Agricultural University or a national research institute. The purpose of these visits was to consult with both producers and specialists about a particular crop or product, so they would be fully knowledgeable about the required production and post-harvest technologies and the possible risks or problems that might be incurred in producing this crop or product.

Facilitating Contracts or Agreements between FIGs and Buyers. In the case of products that have very limited markets, ATMAs worked to secure contracts or purchase agreements between buyers and the FIGs. For example, in the case of medicinal crops grown in the Patna District of Bihar (see: K.M. Singh, et al., 2005), the ATMA identified those companies that actually process specific medicinal and aromatic crops and then determined each company's interest, if any, in purchasing the crop directly from farmers in the district. If a company was interested, then the ATMA would facilitate the negotiation of a contract or purchase agreement between the company and a group of FIGs to produce a specific quantity of the crop, at an agreed upon price and with the production and quality requirements being clearly specified.

Arranging for Training FIG Members. At this step in the process, all FIG members were already familiar with the crop or product to be produced and were aware that they would be able to market the product. Therefore, the members would be highly motivated to learn the details about the production and post-harvest technologies to be used and how to handle specific problems, should they arise. Since the KVKs have specialists (trainers) for most categories of high-value crops and enterprises, in most cases, the KVK would take the lead in organizing a detailed training course or workshop for the FIG members. In many cases, they would involve other specialists in these courses. In some cases, an entrepreneurial farmer, who was already successfully producing this crop or product in another district or state, would be brought in as a resource person or "farmer professor" to share his/her experiences with the FIG members.

Arranging for Planting Materials and Other Inputs. At the outset of producing a new crop or enterprise, FIG members will need assistance in securing suitable planting material, seeds and other inputs required to product a particular crop or product to specification. In some cases, the buyer would provide the correct seeds and then deduct the seed cost at the time the product is delivered and sold to the buyer. In other cases, the ATMA procured the correct seeds and sold the seed to the FIG members at cost.

FIG Members Produce the Crop or Product to Specification. During the first and second year of producing a specific crop or product, the FIG members will need regular supervision and technical support to ensure that the crop or product is produced to specification. Therefore, the appropriate member of the BTT staff in each block would be responsible for monitoring the crop or product and providing technical supervision and support to all FIG members.

FIG Members Harvest, Handle and Deliver the Product. Given the vast range of high-value crops and products being produced under this market-driven extension approach, the marketing of different crops or products varies significantly. In the case of contract production, the post-harvest handling and delivery of the product is fully spelled out so farmers know precisely what is required. In the case of other crops or products, the FIG members themselves, the FIG leader, or a marketing specialist hired by a Farmer Association (of FIGs) may be responsible for arranging the transportation, processing and/or marketing of the product. For example, in one district in Maharashtra, the ATMA set up a supply chain for several groups of fishermen, who organized into FIGs, so they could sell their product directly to a fish exporter in Mumbai, rather than being taken advantage of individually by fish traders. In short, each product or situation may differ; therefore, it is important that the ATMA take the lead in working out the production, post-harvest handling, processing and/or marketing of different products in a manner that will increase the farm income of FIG members and rural employment within the district.

Institutions Responsible for Establishing and Supporting the ATMA Model

The National Institute of Agricultural Extension Management (MANAGE) played an instrumental role in launching the ATMA model in the 28 pilot districts under the NATP. Especially during the first 3–4 years of the project, MANAGE was responsible for training all of the ATMA directors and then for training most of the key district and block-level extension staff. MANAGE assigned one "state consultant," to coordinate these training activities within each participating state and to provide regular technical assistance to each project district. In addition, MANAGE used both print and electronic media, including a web site and video conferencing, to transmit information and training materials to state, district and block-level extension personnel. For example, all of the district level staff need training and experience in new operational methods, such as: how to carry out a PRA to develop a SREP; how to establish representative ATMA GBs at the district level, and FACs in each block. Then, it was necessary to orient these GBs and FACs about their expected responsibilities. Finally, the ATMA and district staff needed to learn new job responsibilities such as procedures for identifying markets, organizing FIGs and then linking these FIGs to these markets, including contract farming. All of these activities reflect totally new skill areas that must be developed among the different cadre of ATMA and extension staff within each district.

The other institution that MANAGE helped establish at the state level, to take over increasing responsibility for these on-going training and technical assistance activities, was the State Agricultural Management and Extension Training Institute (SAMETI) which was established in each of the seven project states under the NATP. SAMETIs are the apex-training institute within each state responsible for extending these institutional and operational innovations to new districts in each state. In each SAMETI, there is a core team of Master Trainers who are expected to carry out "job specific" and "process specific" types of training (as described above) in decentralizing and consolidating the extension system within each district. In summary, these

SAMETIs function as *mini-MANAGEs* as they extended the ATMA model within each state.

On the research side, the most critical institution providing technical support to ATMAs at the district level would be the Krishi Vigyan Kendra (KVK) or multidisciplinary *Farm Science Centers*. The KVK concept began in the mid-1970s and these Kendras are funded by the Indian Council of Agricultural Research (ICAR); most districts in the country now have a KVK, although many still lack adequate human, physical and financial resources. Most KVKs are affiliated with State Agricultural Universities (SAUs), but some are affiliated with ICAR institutes or operated by NGOs. Each of the established KVKs has a well trained staff in field crop production, horticulture, livestock production, agricultural mechanization and home science, plus additional technical expertise as needed within a particular district, such as fisheries, agro-forestry, soil science, or plant protection. This cross-section of expertise enables these Kendras to take a *farming systems approach* in organizing their training and related activities. Also, the KVKs are expected to undertake on-farm testing of technologies developed by the ICAR institutions, and then for training the field extension staff and farmers in these new production and value-added processing technologies.

Achievements and Impacts

The implementation of the ITD component of NATP was monitored and evaluated (M&E) by an independent agency, the Indian Institute of Management (IIM), Lucknow. The resulting M&E reports revealed that these institutional and operational reforms, as outlined above, had been largely achieved. In addition, IIM Lucknow documented the following project impacts:

- More than 10,800 crop or product-based FIGs had been organized at village level, with 85 **Farmers' Associations** (FAs) or **Farmers' Federations** (FFs) being organized at the block and district levels.
- Approximately 700,000 farmers, including over 100,000 women farmers, directly benefited from these new extension programs through a combination of exposure visits, farmer training courses, on-farm trials, demonstrations and so forth. In short, about 15% of all farmers in the ATMA project districts were directly touched by this reorganized extension system.
- More than 250 successful farmer-led, market-based innovations were implemented and documented within the ATMA districts (IIM-Lucknow, 2004b).
- Many ATMAs, such as those in Maharashtra, developed strong partnerships with private sector firms, ranging from poultry marketing; organic farming; the production, processing and marketing of medicinal and aromatic crops; and the export of specific commodities (basmati rice, baby corn, snow peas, etc.); to jointly operating Information Technology (IT) kiosks in collaboration with block-level FIACs.
- Finally, ATMAs have promoted eco-friendly, sustainable agricultural technologies, such as integrated pest management (IPM), Integrated Nutrient Management (INM), organic farming, and the use of water conservation practices, including well recharging, converting from water-intensive crops, such as paddy and wheat, to water extensive crops, such as vegetables, floriculture, maize, oilseeds and pulses. Also, all ATMAs have promoted the use of micro-irrigation systems.

In addition to these institutional and technological achievements, these ATMAs have contributed directly to increased farm income and rural employment through agricultural diversification. For example, IIM-Lucknow empirically documented the following impacts of the ATMA approach on the cropping systems and farm income across the 28 project districts during the four-year period from 1999-2003:

- Horticultural cropping area increased from 12% to 16%
- Oilseed crop area increased from 3% to 11%
- Herbs, medicinal and aromatic crop area increased from 1% to 5%
- Area planted to cereals (wheat and rice) declined from 55% to 47%, but yields increased 14% resulting in no appreciable loss in staple food crop production.
- Average farm income in project districts increased 24% during this four year period, in contrast with only 5% in non-project districts (Tyagi and Verma, 2004).

Conclusion

Decentralizing a large and complex national extension system is not an easy task, but the Government of India appears to be moving toward this long-term goal. Based on the successful pilot-testing of the ATMA model in 28 project districts under the NATP and the introduction of the ATMA approach in 37 additional districts under the Uttar Pradesh–Diversified Agricultural Support Project (UP–DASP), the Government of India is now actively considering plans to scale-up the ATMA approach to 187 new districts or a total of 252 districts in 29 States and 5 Union Territories during the current tenth plan period. To achieve this goal, the Government of India and the World Bank are discussing a National Agricultural Innovation Project (NAIP) that would help finance this planned ATMA expansion. In addition, the World Bank-financed Assam Agriculture Competitiveness Project (AACP), which became effective in 2005, will introduce the ATMA model into 11 districts in Assam in a systematic effort to decentralize its extension system. Next, plans are underway for a second UP-DASP, which will complete the implementation of the ATMA model in all 70 UP districts. Finally, the State of Maharashtra has already initiated actions to implement the ATMA model in all 21 districts of that state. A new World Bank-financed Maharashtra Water Sector Improvement Project (MWSIP) has allocated sufficient resources to complete the implementation of the ATMA model in the remaining 17 districts of that state (in addition to the 4 districts included in NATP).

Final Thoughts

The move from a policy of *food security* to a new strategy that focuses on *agricultural diversification* aimed at increasing farm income and rural employment carries with it implicit risks for the small-scale farm households who are expected to benefit from this approach. The ATMA director and other agricultural leaders within each district need to continually assess their comparative and competitive advantage in producing different high-value crops and products. There will be continuing instances of over production of different commodities and falling prices; these problems cannot be avoided, but they can be mitigated by maintaining a diversified portfolio of commodities, products and enterprises within each district and continuing to seek out new markets and opportunities. The most critical output of this strategy will be that the current generation of farmers will learn new technical, management and organizational skills that will be passed on to the next generation as they seek employment outside of production agriculture.

References

Indian Institute of Management (2004). Lucknow, *Impact Assessment Report*, on the Innovations in Technology Dissemination (ITD) Component of the National Agricultural Technology Project, Agriculture Management Centre.

Indian Institute of Management (2004). Lucknow, *Successful Case Studies*, Interventions and Innovations in Technology Dissemination, Agriculture Management Centre.

National Institute of Agricultural Extension Management (MANAGE), (2004). *Process Change in Agricultural Extension: Experiences under ITD Component of NATP.*

Samanta, R.K. and M.R. Hegde (1998). *Agriculture – Status, Growth and Development, India – 50 Years of Independence: 1947 – 97*, New Delhi, B.R. Pub. Corporation.

Singh, J.P. (2005). *From Self-help Groups to Commodity-based Commodity Associations: The Indian Approach to Mobilizing Rural Women*, presentation at the Workshop on Building New Partnerships in the Global Food Chain, Chicago, June 29-30.

Singh, J.P., B.E. Swanson and K.M. Singh (2005). *Organizing and Linking Farmers with Markets: Experience of the NATP Project in India*, presentation at the 15th Annual World Food & Agribusiness Symposium, Chicago, June 27.

Singh, K.M., B.E. Swanson and J.P. Singh (2005). *Development of Supply Chains for Medicinal Plants: A Case Study Involving the Production of Vinca Rosa by Small Farmers in the Patna District of Bihar India*, paper presented at the Workshop on Building New Partnerships in the Global Food Chain, Chicago, June 29-30.

Swanson, B.E. and P.N. Mathur (2003). *Review of the Agricultural Extension System in India*, unpublished report.

Technology Dissemination Unit and MANAGE (2004). *Project Completion Report*, Innovations in Technology Dissemination Component of the National Agricultural Technology Project, MANAGE.

Tyagi, Y. and Verma, S. (2004). *Economic Rate of Return of Innovations in Technology Dissemination Component* of the National Agricultural Technology Project, submitted to the National Institute of Agricultural Extension Management (MANAGE), Hyderabad.

9

Meeting Farmers' Information Needs in Vietnam

Nguyen Van Linh

Introduction

The Agricultural Diversification program in Vietnam aims to increase competitiveness of agricultural commodities from the country's Central Highland and Coastal regions which continue to have a high incidence of poverty. The program aims to increase farmers' incomes and rural employment by promoting market-oriented agricultural diversification. This would contribute to the growth of the rural economy and the reduction of rural poverty in the region and also broaden the scope of support currently provided by adding a new focus on market competitiveness and quality issues of smallholder's agricultural produce through strengthening information and knowledge systems.

Based on the preliminary findings and other on-going and planned initiatives, a study on farmers' information needs on agricultural production, processing and marketing in Central Vietnam has just been conducted in order to provide inputs to the design of the program, of which possible activities may cover the following areas:

- Production systems to increase the quality of crops and livestock to meet market demand, including capacity building of service delivery institutions (for example extension and veterinary services);
- Farmers' post-harvest technology, including primary processing and storage;
- Marketing of agricultural produce, including supporting farmer's organizations, improving market infrastructure and conducting market studies; and
- Food quality and safety.

The study has been divided into two parts: (i) a review of agricultural information and/or need assessments of small farmers, and (ii) field study on farmers' information needs assessment.

Assessments of Agricultural Information Needs of Small Farmers in Vietnam

This section presents findings from the review of existing literature and past surveys complemented by selected interviews with key informants working for some donor funded projects and agencies under the Ministry of Agriculture and Rural Development (MARD). The purpose of this review is to find out what is known or unknown about farmers' information needs; to identify the critical gaps and the major hypotheses that will be tested through the field survey. The main findings of the literature review will inform the study design (both in terms of content and methodologies, including questionnaire) to fill up the gaps in knowledge.

The Vietnamese government has expressed interest to explore and develop activities to respond to the heightened demand for diverse and better quality agricultural and food products. In order to make informed production and marketing decisions, farmers need timely and relevant information on many aspects of production, such as potential of technologies, market demand and locations, product quality and prices, input availability and sources, processing, credit and even off-farm opportunities, as well as advice on how to make best use of this information.

Farmers receive information from various sources, both from informal and formal sources. Informal information exchange between individual farmers, family members, farmers and local traders, etc. is frequently the most important mode of interaction in the rural communities. In the most remote areas particularly, physical distance, language and social barriers may restrict contacts between poor farmers, the disadvantaged groups (such as women and ethnic minorities) and formal information sources. As Vietnamese farmers are moving away from subsistence farming toward more commercially oriented farming practices, the information needs may change and the importance of formal information sources, such as farmers' organizations, government agencies, agricultural extension, plant protection and veterinary services, banking sector, and the media, is likely to increase.

A number of needs assessments have been carried out by the government, public institutions and international donor funded projects in Vietnam, especially in the Central Highland region. However, farmers feel that information on agricultural production, processing and marketing issues is lacking.

More detailed assessment of farmers information needs is required to prioritize appropriate interventions that would allow farmers to access information and plan and diversify the quantity and quality of production in line with market demands.

Findings from various efforts

There are presently several government agencies and donor-funded initiatives working in the field related to the provision of information, whether it is in gathering, analyzing, or disseminating it.

Beside the ministry, other government agencies and non-government organizations (both local and international) play an active role in supplying farmers with information, including : Farmers Union, VACVINA (Vietnam Gardening Association), Bee-Keeping Union and universities.

Informatics Center for Agriculture and Rural Development (ICARD)

The Market Information Division of MARD collects market information from various sources (including external information agencies, i.e. *Reuters*) and provides updated information on the price of main commodities/goods in the some local and world market. This provides information on both international and domestic markets of the 10 key products of Vietnam. There is price information and also articles analyzing the market trends of specific products. The production information section provides typical and progressive production models throughout the country. The science and technology section provides new farming techniques of various key products. It also provides lists of new seeds and animal breeds.

Recently, ICARD has provided the following publications:

- Bulletin for Leadership: (1) providing information on strategies, policies on agriculture and rural development of other countries in the region and the world; (2) drawing experiences and lessons regarding organization, management, guidance and policy mechanism; (3) analyzing and forecasting tendencies in the development of agricultural production and business.
- Bulletin on Production and Marketing for Agriculture and Rural Development: updating information on production and business situation; products' supply, demand and prices; disasters, diseases and new agricultural policies.

Trade Promotion Program in MARD

The development of market information system in provinces is one of the activities of the Trade Promotion Program under MARD (ICARD as the implementing agency). Before 2002, this was a pilot project funded by the Government of Sweden. The project aimed at developing a system to collect market price of certain products in Phu Tho province. People were contracted to go to the market to get market price of certain products. The market price was then edited into bulletins or displayed in the notice board at the market. An evaluation, therefore, revealed a great demand for market information among the farmers.

The system was expanded to 7 provinces. In these provinces, the program provided budget for training of agriculture staff on how to use Internet and how to collect and process market intelligence. The staff was then contracted to go to the market or enterprises to get market price of specific products in each province.

The information was then transmitted to the persons in charge in the Provincial Department of Agriculture and Rural development (DARD). And here, information was processed and uploaded to the website *www.agroviet.gov.vn*.

Because the Trade Promotion Program provides budget only for training, the operation of the entire system depends on the ability to mobilize provincial budget. Only in 2 provinces this market information system worked well. There the coordination with local newspapers and television stations was also established to disseminate the collected information.

Since October 2003, the market information system in ICARD collects price information in 11 representative pilots in the South and 4 in the North. However, the existing market information system (as pilot project) focuses only in some kind of fruits and vegetables and does not be processed and analyzed. This source also needs to be disseminated more widely through various media (newspaper, TV and radio...) as well as other relevant tools/channels in order to reach wider range of recipients, especially farmers. In addition, the cooperation with other information channels such as the agricultural extension network, farmers' associations, commodity associations, farm clubs, etc. to provide farmers with technical information on their production, processing is not yet enhanced. From the 1st Jan 2005, ICARD is oriented towards informatics activities only, and market information research and trade promotion activities will be shifted to the Institute of Agricultural Economics/IAE. The working group of ICARD in the South operating the fruit and vegetable information system will now be contracted with the new institute. Such activities of validation, processing and dissemination of agricultural information are also linked closely with the new functions of the institute. Other activities like information on quality, equity standards, and institutional support for production process and market promotion will also be based firmly on research activities assigned for the institute.

National Agricultural Extension Center (NAEC)

The NAEC has been established by MARD in 2003. One of NAEC duties is to support Vietnamese farmers in poverty reduction, enhance competitiveness of their main products, transfer of new technologies, and provide training programs and documentation (e.g., leaflets, video tapes, CD rom, etc.) to the extension staff and farmers. With respect to assessment of farmers' information need, there was not any study carried out, and most of designed projects on extension are based on information gathered from provincial reports and suggestions. Technical books on specific new crops or animals have been published (and provided free of charge among the extension network) mostly based on recommendations from scientists and/or government policy-makers.

Information Capacity Strengthening for Agricultural Policy Formulation (MISPA) Project

The MISPA project was designed jointly by MARD and the French Ministry of Foreign Affairs to strengthen the capacities of ICARD to achieve its mandate. The Project consists of three closely linked components : ICARD capacity building; implementation of specific studies on agriculture and rural development; and project management.

The first component supports all activities carried out by ICARD, with special emphasis on market information, information technology, and information publications. It aims to provide ICARD with assistance on: (i) strategic management and equipment improvement; (ii) development of skills for socio-economic information analysis; and (iii) information dissemination. The second component comprises a study fund and three thematic working groups. The study fund finances sector or thematic studies needed for relevant strategic decision and policy formulation. The studies are implemented by specific teams of experts identified in ICARD and among its partners. This component is open to any financial contribution of external donors. The three thematic groups include: (a) competitiveness of main Vietnamese agricultural commodities; (b) rural development strategy; and (c) perspectives for the coffee sub-sector.

A CIRAD report on "*Setting vegetable market information systems in Vietnam*" was presented to ICARD/MISPA project with specific objectives to inform farmers, traders, development and policy agents on possible new, untapped, profitable outlets for vegetable products; to have private and public stakeholders share common vision on problems and solutions (technical, institutional) to gain market shares. The report also recommended that the market information system should adapt to diversity of users and needs; adapt frequency of collection and dissemination to price variability; seek homogeneity of products, time of collection, units of sale; combine information with stakeholders' consultation (market information and conciliation systems) to deal with expressed constraints other than information (technologies, access to inputs and transport, etc.).

IFPRI study on "Fruit and Vegetables in Vietnam – Adding Value from Farmers to Consumers"

This is the report for the project "Development of Post-harvest Activities and Agro-industry as a Strategy to Improve Rural Livelihoods in Vietnam", implemented by the International Food Policy Research Institute (IFPRI) in collaboration with Hannover University, the Ministry of Agriculture and Rural Development (MARD) and other Vietnamese institutions. One of the study objectives was to identify and characterize institutional mechanisms involved in lowering transaction costs and to increase the access of rural households to information, markets and assets.

The project carried out three national surveys of different participants in the fruit and vegetable sub-sector: IFPRI-MARD survey of Fruit and Vegetable Producers with a sample of 1,505 commercial growers in 21 provinces carried out in Oct-Nov 2000; IFPRI-MARD survey of Fruit and Vegetable Traders with a sample of 110 traders and exporters carried out between May-August 2001; and IFPRI-MARD survey of Fruit and Vegetable Processors with a sample of 241 processors carried out in 2001. In addition to three surveys mentioned, the study team analyzed data from two Vietnam Living Standards Surveys carried out in 1992-93 and 1998 and conducted two small informal surveys of institutions (June 2001) interviewing key informants from research institutes and MARD departments and international consultants, retail outlets in Hanoi (also in 2001).

Findings from the survey of fruit and vegetable producers showed that access to communication is important in the marketing of agricultural products. Very few producers had a telephone, fax machine and none used email. Producers receive market information from a variety of sources, including extension officers, banks, traders and official government sources. Most producers obtain price and market information and information about regulations from personal contacts, or word of mouth, some from the radio or television, cooperatives and traders. Credit information comes overwhelmingly from banks. Information about new varieties usually comes from personal contacts with extension agents. This follows for technical advice as well, whereas management advice comes from the people's committee (and also personal contacts). Overall, most of the survey respondents received information from extension service department (63%) and that majority of the advice was in the form of information on fertilizer use, pest management and varietal selection.

The survey report suggested that as fruit and vegetable marketing extents over longer distance, the need for timely and accurate information about prices and market conditions grows. Market information services must focus on key products and markets to avoid over-extension. Furthermore, they must incorporate regular feedback from users to ensure that they remain useful and relevant.

Farmer's need study (PAR/MARD)

The project VIE/98/004 in MARD, co-financed by UNDP and Royal Netherlands Government, is aimed at supporting the ministry to strengthen its institutional and human resource capacities to support Public Administration Reform (PAR). A farmer needs assessment was carried out to provide important information on how to orientate the restructuring process of MARD towards the field level and which role MARD should play in service delivery to farmers, livestock keepers, foresters and other stakeholders.

The needs assessment consisted of three consecutive phases: review of existing documents and information on farmers' problems and needs and development of a draft interview guide; motivational (qualitative using semi-structured questionnaire) research on farmer problems and needs regarding public service delivery covered by MARD with a sample of 192 farmers in eight provinces across the country; and quantitative survey, using structured questionnaire for validation of qualitative results with a number of 1,261 interviewed farmers (belonging to 30 districts and 60 communes). Both phases were completed in 2002.

There is a general feeling of confusion among farmers regarding the quality of the variety of products available on the market. Most farmers are not able to find market opportunities and are dependent on middlemen who dictate the price. On the other hand, farmers feel so dependent on middlemen that they are not able to negotiate the prices even if they are informed about the current market prices. In some areas, farmers can sell their products to processing factories or to industrial slaughterhouses, but problems arise due to unclear contracts and unstable prices. Farmers need more information on price of inputs and agricultural products and how to sell their products, there must be standardization of the method to evaluate product quality. They also need more information regarding credit sources.

TV programs on agricultural technologies are useful, but not specific enough to the different climate and local conditions, and not applicable in the field. It is proposed that agricultural public service providers should make more use of the local TV stations for transferring information and techniques to farmers. Extension agents should use videotapes when explaining new techniques to farmers. Radio programs are very useful, but the information is still poor. Farmers would like to buy technical and practical books on agricultural techniques with photographs and images. For farmers information on output price and technical information is important. They want to have an extension program in a local language in radio/TV.

The study also recommended that the loudspeaker and village head are very useful information sources that should be leveraged nationwide and specially focused in areas where the TV penetration is low. In addition, TV broadcast could also be developed as one of the key sources, because of its wide coverage and the ability to transmit both images and sound (visualization would enhance the effectiveness of the information provided). Information contents should be practical and specific and preferably in the local language.

Project Support to Public Administration Reform Program in Ministry of Agriculture and Rural Development VIE 02/016: FEASIBILITY STUDY IN SUPPORT OF RURAL PILOT TELECENTRES FOR FARMING HOUSEHOLDS IN VIETNAM

The feasibility study took the form of a structured field survey of 20 selected communes and a review of national and international experiences with rural Information and Communication Technologies (ICTs). The purpose of the field survey was to understand the needs of communities as a market research on farmers' demand for communication and the needs of information, as stated by the recipients themselves.

Findings from the study showed that in all communities, the biggest difficulty encountered by farmers is the lack of market information for their outputs. In both those communities where the farmers' target is the domestic market, and communities with products for exports, farmers sell their products through individual middlemen. Processing enterprises rarely come to collect raw products or contract farmers in growing specific crops and commit to purchase qualified outputs.

There are various perceptions among farmers on the usefulness of information on price of agricultural products. In some communities, farmers do not show much interest in market price information, simply because they cannot take their products to the market place, but sells through middlemen. The fact was that farmers were not able to take advantages in negotiating for a higher price in these cases because they were not aware of market price information.

The study also found that systems of agricultural extension, veterinary and plant protection are the most important sources providing technical information to farmers. However, capacities of commune extension workers in providing farmers with updated information are still limited. Every year, communes' extension workers attend 1 to 2 training courses, held by district or provincial extension agencies, but they are provided with only some information on new seeds, breeds, farming techniques. They are not equipped with the methodology and skills in working with farmers and thus, they seemingly cannot provide training to farmers. In addition, their capability of collecting and compiling information from different sources is also limited. In most communities, what extension workers can do is to receive information from the district agencies and orally disseminate such information to the farmers mainly through loudspeakers or social organizations to the district agencies.

Generally, technical information is currently disseminated to the farmers mainly through extension training courses. However, participants of training courses are mostly better-off farmers. The poorer, who have little land and few other production materials, rarely attend because they are not interested in applying new techniques which require considerable higher investments than traditional ones, or they may choose to live on daily wages earned from working for those who have more land.

The development of demonstration models is a good way to transfer information and technology to farmers, especially in remote and/or ethnic minority areas, who prefer to see realistic positive results before deciding to apply a new technique, new crop variety or animal breed. Agricultural extension stations at district level now play a major role in the development of demonstration models, but stop at technical aspects and do not provide any assistance to marketing of their products.

Conclusions and implication for the field study

As mentioned in previous section, findings of the initial literature review as the first part of the study aims to find out what is known or unknown about farmers' information needs, and to identify the major hypotheses that will be tested through the field survey. The main findings of the literature review are presented below with implication for the design of the guide and questionnaire for focus group discussions and face-to-face interviews in the selected communes.

1. The initial literature review showed that market information on output prices is of most concern for farmers. Although many efforts have been made by various government agencies and donor-funded projects, using different delivery tools to provide this kind of information, many farmers have not got

access to updated price information in order to negotiate with the buyers. Other aspects of market information, such as input prices and locations, market demands on volume and quality/standards of products, packaging, storing and transportation conditions, etc. were not (or not much) mentioned in the previous studies yet.

2. Farmers from different categories of gender, age, ethnic groups, farm size, wealthy/income level, main livelihood source, education level, etc. and in various geographical areas may have their own needs on information. These concerns although mentioned in the previous study designs, but have not been clearly analyzed and indicated yet.
3. One of the questions which has not been answered yet is who is providing the sources of information with information on which topics are seen as the most appreciate/confident by farmers? It seems that most of information sources are trying to provide information on wide range of topics, but nobody knows how useful this information is to (different groups of) farmers.
4. With respects to technical information, such as production techniques (including new seed/crop varieties, animal breeds and new farming practices), veterinary and plant protection, etc. most attention is paid to the provision of that kind of information to producers as much as possible (through extension and other services). Training and demonstrations provided by public extension agencies (including veterinary and plant protection) seems to be more appropriate to farmers in improving their technical knowledge and skills, but we need to know how many farmers (of different groups) can get the access and to which extent they can understand and use/apply these information.

5. It was also not very clear that, in farmer's opinions, what are the most important and appropriate interventions improving farmers' access and use of appropriate information, why and how the farmer can prioritize their suggestions, how farmers can use the information in changing their livelihood, etc.

Another issue regarding to sustainable aspect of any intervention is to improve farmers' access to information. The question is for what kind of information services are farmers willing to pay needs also to be verified.

IDENTIFICATION OF FARMER'S INFORMATION NEEDS

The study

As mentioned earlier, there have been a number studies on farmers' needs carried out by research institutes, NGOs, national or international projects/programs, etc. in Vietnam. However, information on the farmers' needs are scattered and restricted to some specific areas/issues that such projects/programs focus on.

The objective of the Assessment of Farmers' Information Needs is to identify the current availability of, as well as the potential unmet need for information supporting farmers' agricultural production, processing and marketing activities, and thereby advise the Government of Vietnam on priority interventions to address the identified information needs of farmers. Specifically, the assessment focused on the following four questions:

1. What are the information needs of farmers (women, men, ethnic groups, etc) on agricultural production, processing and marketing?

2. What are the information sources of different farmers, for various topics?
3. To what extent can different farmers apply the information available?
4. What are the most important and appropriate interventions that improve farmers' access to and use of appropriate information?

A survey proposal was developed with the structured questionnaire for face-to-face interviews, and guidelines for focus group discussions were conducted with farmers, relevant organizations (e.g. Farmers' Unions (FUs), Cooperatives, Women's Unions) and government bodies (e.g. provincial/district DARDs and Extension Services, Veterinary Services, etc). The following aspects have been included:

1. Information needed for production and farm management:
   - To make planning decisions, including information on e.g. government production policies, market locations and demand, prices, quality requirements, inputs, weather, etc.
   - Farming practices, such as production and post-harvest technologies, veterinary services, inputs and irrigation, etc.
   - Farm-enterprise management, credit and other financial issues.
2. Farmers' current information sources:
   - Formal and informal contact persons/ organizations (e.g. individual farmers and traders, farmers' organizations, commune and district service providers, etc)
   - Information sources by topic, extent and satisfaction with quality and availability of information.

3. Constraints/opportunities related to using information and information sources, including:
   - The capacity of the farmers to meaningfully to use information (in the form of questions addressing ability to apply available market information, technical information, etc.).
   - Farmers' access to information, e.g. type of media, timing, location, etc.
   - Format of information.
4. Information needs not currently being addressed.

Gender sensitivity has been included with the above topics during the interview with farmers. The study conducted separate assessments in three provinces, representing remote mountainous areas, coastal areas, and poor peri-urban areas, respectively. The areas selected in each of the three provinces are representative of all sections of the farming community (one commune per district, a total of three districts per province).

The focus group discussions and face-to-face interviews were conducted with commune' Farmers' Unions, Cooperatives, Women's Unions, VACVINA and government bodies, such as provincial and district department of Agriculture and Rural Development, provincial Extension Center, district extension Station, provincial and district Veterinary Services, provincial and district Plant Protection Services, etc.

The face-to-face interviews and focus group discussions were carried out with farmers engaged in crop production, animal husbandry, aquaculture, processing, forestry or combination of these activities. Interviews and discussions comprised both men and women from different age groups (e.g., <35, 36-55, and >55), having different socio-economic conditions (better off, middle, and poor) and belonging to different ethnic groups. Consideration was given to the case of women-headed households.

The wealth of the household is important selection criteria, but another aspect is also the wealth of the district, commune or village. The selection of districts and communes for the field survey was based on characteristics of households, such as main livelihood sources, types of farming activities (i.e., agriculture, aquaculture, agro-forestry), income levels (poor, better-off communes), ethnic groups, farm sizes, education level, etc. with attentions to issues of geographical coverage (mountainous areas, coastal areas, and poor peri-urban areas where applicable). In this sense, nine communes in different nine districts were selected.

The interviews in each province covered 500 farming households (making a total of 1,500 individual interviews) and took into account the specific needs and priorities of different categories of farmers, including women and ethnic groups. Although the surveys have special attention to ethnic minority households, they did not limit to the ethnic minority households only, but also cover Kinh (majority group in Vietnam) households.

In each selected commune, a sample of around 170 individual farmers was randomly taken from the commune list by the systematic sampling method. The existing wealth data of farmers' categories (households divided into poor, medium and better-off groups) at the commune and village levels were used and random sampling activities were done within each wealth group (in order to guarantee a representative sample across the wealth groups in a given commune).

Face-to-face interviews using structured questionnaire were undertaken at village level. This activity aimed at collecting opinions, motivation, attitudes, and perception of farmers' situations and their information needs on agricultural production, processing and marketing of produce as mentioned in previous section.

Findings from the field

Findings from the visited areas showed that, generally, farmers in the rural areas receive a variety of incoherent information from many different sources. With respects to the rural livelihood, there are two main types of information at commune and village levels, consisting of technical and market information (other types of information such as government policies, weather, irrigation, etc are not being discussed below).

The technical information can be understood as technical guidelines on agricultural, forestry and aquaculture production (including processing/post-harvest activities) that farmers could receive from the existing extension system (both public and volunteer/private extension) or others. In many cases, technical information has been also delivered by communal officers and/or village leaders, especially in the remote mountainous communes/villages. Mass organizations (FU, WU, etc), input supplying companies, and individual traders (being seen as private sector) are also providing farmers some related technical information in an informal basic. The mass media (such as local loudspeaker, TV, radio systems and newspapers, magazine, etc) play an important role in providing farmers technical information, but farmers' access to and their ability in understanding of information messages from these sources does not always exist. One may distinguish two main sources: public (extension, administrative system, communication agencies) as the *formal* source and mass organizations/private sector as the *informal* source. Each of these two sources of technical information plays their own roles in the rural community, but the first (public) is dominating in this field.

Another concern is the design of the messages. In the visited areas, the study team found that there was a gap between central and local sources and between providers *per se* in designing technical information messages. In many cases, the public source of information (especially communication agencies) was providing technical information in general, but not be based on or modified to fit specific local conditions. This situation calls for a local agency/agent who can select, consolidate and adjust the information before providing villagers with better understandable and applicable.

The availability of technical information is limited in remote/mountainous communes, especially at the village level. In addition, poor villagers/farmers with low level of education could not been involved in extension activities as they might be busy with their field-work all the time, or lack communication means in order to get access to information sources. Making technical information available and accessible at commune and village levels would be a good strategy of the public sector.

Training courses, demonstrations/models, village/ commune meetings and loudspeaker system could be seen as more effective approaches/methods providing technical information to farmers directly than other means. Given farmers' limited abilities to get access and understand the technical message, these approaches could regularly provide farmers with information in a simple way using local language and making them more confidences to the result if they would like to apply a recommendation. While visual media such as video tapes/CDs and pictures could help farmers in realizing farming practices and/or more specifically classifying/characterizing types of pests/diseases or crop's growing stages, the local loudspeaker system is frequently providing and reminding them information on timing and numbering (for instance, time and quantity of fertilizers to be applied in the field) using a simple way of communication. However, without carefully selection and editing information would create confusion and troubles for farmers in choosing which recommendation they should apply.

For a long time, the public extension has provided farmers with the technical information they have, but not much with the information which farmers feel they need. There are many reasons for the situation, including limited fund, lack of extension staff, poor working conditions, etc. In addition, farmers lack collective voice. Individual farmers may have so many different (information) needs that the public agencies could not satisfy all of them, while the private service is still weak or not present in this field. This leads farmers to become passive recipients of information rather than be actively involved in the communication process through which they can express their needs, participate in seeking and selecting relevant information, etc. In all visited communes, cooperatives are still weak in terms of its functional and operational aspects. Strengthening of the existing extension system according to local needs (using participatory approach) could be an option to make the technical information available to farmers.

Recently, *the market information* has been broadcasted nationwide through mass media (TV, radio and newspapers). MARD and Ministry of Trade (MoT) are lead agencies providing price and also some other related market information on daily, weekly and monthly basis. Although farmers see price as very important information, they are not satisfied with what they received from the public agencies. This information is often too general/not specific enough for given products or commodities with dubious instructions of qualities and standards required. Getting incorrect or out-of-date price information constrains farmers in negotiating with dealers. In all visited communes, the study team did not see any evidence of local agencies providing detailed information on market demands (including prices) that are useful to farmers. The interviewed farmers expressed that they often rely on information got from the dealer and/or their neighbors. The dealer is seen as middleman who plays a very active role in the rural area. Farmers are often convinced that the real market information is on the dealer's hand. One of the study team concerns is that most of interviewed farmers as producers have paid more attention to market prices, but less to other market demands (such as quality and standards of the commodities, market place, packing, transportation conditions, etc). This issue may result in a lack of knowledge on market demands and farmers would not be seeking the way to improve farming/processing practices for a better quality and then better price of their products.

In the communication process, poor linkages and one-way communication from the district down to the commune and village level is a constraint. Providing the e-link could be seen as one of instruments supporting solutions for this problem when a communication point at the commune level is being established (it maybe called as the commune information center (CIC) - see below recommendation). In the remote and mountainous communes without telephone line, this instrument seems to be unrealistic or impossible at this time. In that case, an inter-commune communication point could be a solution (this e-link is installed only in the case where one of two or more communes are connected to the telephone line).

The private sector (firms/companies or individuals) in the visited communes is still weak in terms of small business levels. As mentioned above, individual businessmen (middlemen) have just come to villages/communes to sell material inputs and/or buy farm products at a small scale without signing any contract in advance. Farmers may get some market information from their business partners. In the future, the private sector (individuals or firms) may participate in this process as one of market information providers by collaborating with the commune information center, which will benefit them and farmers in the area.

Recommendations

The development of Commune Information Centre (CIC)

According to the provincial record nearly 50% of the communes in the three provinces we visited have a post-office (the rest are communes in remote/mountainous areas). This office has function for telephone and fax services, receiving newspapers, magazines, letters, documents, etc. from outside then circulating them to the commune people. It is clear that the existing rural post-offices could not be able to act as a communication center serving farmers with the information they need because it lacks the necessary equipments, and has only one local staff.

Discussions with officers from the provincial and district extension agencies and information gathered from interviewed farmers recommended that a Commune Information Center should be established to make relevant information available to farmers. The pilot center in selected communes could be operated by using the existing commune post office. This proposal is in accordance with MARD' 5-year plan for the period of 2006-2010 showing that *100% of all communes have a post office, which gives access to technical and market information to rural people* (MARD, 2005). Both technical (production and post-harvest) and market (prices, quality requirement, etc.) information can be provided to farmers in the commune through this center and village loudspeaker system.

The question is how to equip and operate this center in an effective way? Who will support the center in its operation? Discussions with local extension agencies in visited provinces and districts showed that the commune information center could be connected with district information providers such as extension, plant protection and veterinary officers who may act as collaborators or consultants for technical assistance to the center in this field.

Regarding local information sources, the formation of a Coordination Unit (could be inside provincial extension center) to coordinate all kinds of information sources (including information from MARD system), verify, re-format, update and provide them to the CIC within the province daily/weekly or monthly should be made. For the provisions of *technical information*, the existing extension system (including district and provincial veterinary and plant protection stations/ centers) can be a good provider. There is still a question of who can provide farmers with preferred *market information*? Further discussions with the provincial Trade Department should be conducted to find out appropriate options. Other option may be the use of market information network under MARD, that the Institute of Agricultural Economics (IAE) is now establishing.

Farmers in the commune can come directly to the CIC to get information needed, or the CIC can provide them through village loudspeaker system where available (it also depends on the type of information they may need). The CIC would also be responsible in collecting farmers' information needs and then send them to the relevant providers. This also implies farmers or their organizations will have opportunity to participate in the process expressing their demands on information and providing the provider with feedback on types, content or format of the message they got.

Capacity building activities for this purpose are needed, and this system requires an Internet connection, computers, telephone, etc. A staff working as information agent at the CIC could be added. Here will be an issue of human resource. It is proposed that the commune extension worker would be another option.

Stimulating demands of the farmers

As mentioned above, findings from the field showed that farmers sometime neglect other aspects of market information. They see the price information as the most important, but give less attention to other market requirements such as type, quality/standard, package, timing, location, transportation, etc. for their commodities. Demands of the farmers must be stimulated through a combination of different tools, including community mobilization through the network of farmers' cooperatives/associations, collaborators, close links with social organizations in the communities (FU, WU, etc). Stimulating demands for various services of the farmers not only targets at generating more revenues for the CIC, but also at making the farmers aware of new technologies and market demands, so that their intellectual level is gradually raised. The CIC will function as a two-way communication agent.

Strengthening the extension system and MARD information network to provide strong supports to the CIC

The National Agriculture Extension Centre (and its agencies at local levels) and Institute of Agricultural Economics are the two central agencies under MARD, which are the most closely linked with the future CIC in communes. Currently, the NAEC focuses its activities in supporting farmers in technical side and the IAE in market side. If NAEC and IAE are deeply involved in the operation of communes' CIC, there will be mutual benefits for both the CIC, which means the farmers, and these agencies themselves.

For the benefits of the CIC and farmers:

The CIC and the farmers will be assisted with useful and updated information. Regarding technical information, one of the biggest problems is that information is mostly in a format which cannot be read and understood well by the farmers, especially ethnic minorities. With the assistance of the NAEC system at local (provincial/district) levels, such information can be translated into more easy to understand and easy to apply formats. Regarding market information, public information providers are mostly not available, and if available their information is not updated and analyzed. With assistance of the IAE, which is now running and/

or involving in several programs/projects to set up market information system (MIS), the CIC and farmers will have greater opportunities to get access to the most updated information. In addition, the IAE can help analyzing the information to show the market trends, which provide sound foundation for the farmers to decide whether to farm a particular product. If the provincial *Coordination Unit* is established, the NAEC and IAE can provide them with latest information in both technical and marketing aspects.

For the benefits of the NAEC and IAE:

One of the tasks of these two agencies is to provide good information for the farmers. However, it is very difficult, if not impossible, for these two agencies to rise up their hands to individual communes. Currently, in most cases the NAEC can only reach district level, while IAE can only upload market information to the website. There is yet to be any mechanism for these two agencies to have direct contact with the communities in the most efficient way. The CIC, if being set up, will play the role of an e-link between these central agencies and the communities. With assistance of the CIC, NAEC and IAE can provide their assistance down to the communities at the least cost and human resource.

It is strongly recommended that if the CIC are set up, MARD should assign specific tasks to these agencies and provide relevant assistance to build up capacity of these agencies so that they are capable to play the role of central information coordination unit.

References

Annual Report (2003): *Tea and Fruit Development project VIE 1781 (SF).* This report has been agreed in the Annual Review Workshop 2003 in Tien Giang province from 25-27 February 2004.

Development Associates A/S, (2004). *Review of Agricultural Market Information Systems in Vietnam*. WB document.

International Food Policy Research Institute (IFPRI), (2002). *Fruit and Vegetables in Vietnam: Adding Value from Farmer to Consumer*. Prepared for Deutsche Gesellschaft fur Technische Zusammenarbeit (GTZ) and the Bundesministerium fur Wirtschaftliche Zusammenarbeit (BMZ).

Le Thi Phi, Nguyen Van Duong, Nguyen Ngoc Quang and Phan Lac Vang, with Elaine Morrison and Sonja Vermeulen (eds), (2004). *Making the most of market chains: challenges for small-scale farmers and traders in upland Vietnam*. Small and Medium Forest Enterprises Series No. 2. International Institute for Environment and Development, London, UK.

MARD, (2004). *Feasibility Study in Support of Rural Pilot Telecenters for Farming Households in Vietnam*. Project Support to Public Administration Reform Program in MARD VIE 02/016. Hanoi, April 2004.

MARD, (2005). *The 5-year plan for agriculture and rural sector, period 2006-2010* (Second draft). Hanoi.

Moustier P., Dao T. A. & M. Figuie, (2003). *Changes of the food sector in Vietnam*. Presentation paper to MALICA seminar on the food markets. Hanoi.

Moustier P., (2003). *Setting vegetable market information systems in Vietnam*. CIRAD presentation paper to ICARD/MISPA.

PAR/MARD Project VIE/98/004/B01/99, (2003). *Farmer Needs Study*. Statistical Publishing House, Hanoi.

10

Privatization of Government Agricultural Extension Agencies

M. Kalim Qamar

Background

Agricultural extension started as a formal organization about 50 years ago in most of the Asian countries. As the extension organization was created within the Ministries of Agriculture, which were responsible for agricultural development, the performance of agricultural extension function became a sort of monopoly of the government in the Asian region. In some countries, extension organizations expanded considerably comprising thousands of field extension workers and other staff.

So many political, economic, social and climatic changes have occurred since the formal inception of extension. These days, the calls for privatization of extension services are getting louder, some coming from the donor community. The donors have started pressing the governments for looking alternate modes of extension delivery which would alleviate fiscal and manpower burden caused by public extension services.

Of course, the total privatization of extension as happened in developed countries like The Netherlands, England, and Denmark, has not yet been introduced in any developing country. Extension has been privatized in several Latin American countries as well. Some countries like Estonia, Nicaragua and Israel, have a mixed bag – some public and some private extension advisory services. Interestingly, USA still maintains a large public Cooperative Extension Service although its farmers also get advice from commercial companies.

Why Think about Privatising Extension?

The move for privatising agricultural extension services has not started overnight; it has taken years to develop and pick up momentum. The following will help in comprehending why, in the first place, the thinking about privatisation of government extension agencies started.

Dissatisfaction with public extension services

If subsistence and commercial farmers were happy with the extension support they receive free of charge from the government extension workers, they could have constituted an indeed strong lobby for uninterrupted continuation of public extension services. However, this is not the case. The pre-service academic preparation of the staff for performing extension tasks has always been weak. Working conditions in the profession have been very poor as compared to some other disciplines like agricultural research, and the infamous fragile linkages between extension and research have refused to improve, persisting stubbornly even today. The extension workers, in spite of their good intentions, have therefore, not been able to fully satisfy the extension needs of farmers, making most public services a target of criticism, down-sizing by the government and, in some case, total abolishment.

Entry of other actors in extension

Partly because of general dissatisfaction with the government-run extension services and partly due to several global developments and rising opportunities and desire among well-off persons and institutions for undertaking humanitarian work as well as for starting agri-businesses, several actors have now entered in the extension arena. They include religious and philanthropic bodies, NGOs, private extension advisory companies, individual advisors, farmers' associations, companies dealing in farm inputs and credit, universities and research organizations. In some cases, these actors are delivering extension advice all by themselves while in others, they are following public-private partnership pattern. Some of these actors, notably charity and religious ones are serving farmers in those remote areas where public extension agents are rarely seen mainly due to lack of transportation. The years old monopoly of the government-run extension services is under threat and disappearing gradually.

Dwindling public funding for extension

Public agricultural extension departments rarely had adequate government funding. In some developing countries, these services depended heavily on funds coming from donor-financed projects, even to purchase simple stationary items. Major international financial institutions and development agencies, while prescribing measures to the developing countries for strengthening their ailing economies earlier recommended structural adjustment and, later, assigned low priority to agriculture sector until recently. In many countries, public extension services were drastically downsized, recruitment to fill in extension vacancies was frozen for several years, such as in the Philippines, and already inadequate public funding for extension was further curtailed. In Indonesia, Philippines, Ghana, Uganda, Tanzania, and several South American countries where decentralization took place, extension was marginalized in terms of priority for budget allocation and human resources. The extension organizations, which have been under critique for quite some time, are now struggling for survival in many countries.

Donors' pressure for privatizing extension

Donors have been advising governments to take measures towards cost-recovery, outsourcing and partial or full privatization of agricultural extension services. In several Latin American countries, public extension services have been abolished and alternate modalities of privatized extension and advisory services have been introduced, some being managed by farmers' associations. It is a different issue whether they are functioning better than the now defunct public services. In Uganda, several major donors are financing a project under which National Agricultural Advisory Service is gradually privatizing the entire public extension service. Only time will ascertain its sustainability, i.e. whether the farmers' associations will be willing to pay for the advisory services from their own pocket after the donors' funds regularly received by them from the district governments stop coming in after the project comes to an end.

Eagerness for potential profits in agri-business

There is a show of keen interest in many countries by the private sector to tap the potential profits by engaging in agri-business ventures. The governments in all developing countries are anxious to see their farmers, whether small or large holders, running their farms like business enterprise. However, most small farmers do not know the basics of the trade. The government extension workers, unfortunately, have not been trained to handle such challenges. The situation, therefore, offers an attractive opportunity to those who are either interested in starting agri-business or are eager to sell various types of machinery, technical know-how and raw material to farmers who would like to start such businesses. Global market liberalization have also spurred this desire and many developing countries which enjoy comparative advantage in producing certain agricultural products are willing to initiate private extension advisory support exclusively for the producers of relevant crops and fruits.

Satisfactory private advisory services in developed countries

In many developed countries such as The Netherlands, England, Denmark, Australia, the private advisory services are doing really well. Each of these services has hundreds if not thousands of member farmers who pay fee on regular basis in return for excellent benefits and advice which enhance their income. In Northern Italy, for example, a private advisory service, which gives extension advice to its member producers on quality production of apple, keeps a close watch on local weather and issues alerts on a specific radio frequency to its members about any major risk of frost or some other weather-related calamity so that appropriate precautions could be taken. In Denmark, the Danish Agricultural Advisory Service enjoys the membership of thousands of farmers already for a century. The farmers pay the salaries of its private advisors and get the most profit-potential farm business plans, tax returns and account sheets prepared by the advisors. Many of these countries, where private advisory services are located, are in fact prominent bilateral donors and, no surprise, recommend privatization of extension in the developing countries because the privatization has done so well for their farmers.

Worldwide trend towards privatization

There is general belief that the private sector is more efficient and cost effective than the government in running certain public services. The private companies are normally not as large as some government bureaucracies which are unable to move fast due to their own weight. They also have necessary resources and use better technology to provide a quality and timely service at lower cost. Of course, profit-making is the ultimate motive of the private companies. Sectors like communication, railways, industries, education and health have been privatized in many developed countries. The trend for privatization is now expanding in less developed countries as well. The areas like agricultural extension, which involve long-term investment in human behavioural change, are not that attractive to the private sector as are the sectors which give quick and positive returns to the investment. Still, there are components within agricultural extension such as sale and advice on farm inputs like seed, fertilizer, and pesticide, which could promise almost instant gain of profit.

Undetermined Status of Private Extension in Asia

Unfortunately, there are no reliable data available on the status of privatized extension services in Asia. In 2002, when FAO Regional Office for Asia and the Pacific held regional Expert Consultation on Research, Extension, Research-Extension-Farmer Interface and Technology Transfer, with participation of several Asian countries, none of the country papers carried any section on privatized extension. It does not mean, however, that private extension does not exist in the region. In most of these countries, large cooperatives and commercial farmers have been "buying" extension advice from private sources in a rather ad hoc manner. Some large, commercial cooperatives have been said to have their own extension advisors. But the presence of a well organized and established private extension advisory service has not been heard in Asian countries which are not yet as developed as Japan or South Korea.

A comprehensive study is indeed needed for determining the extent and type of private extension advisory services that presently exist in various Asian countries. The findings of the study will be of great value in determining the future course of action on extension privatization.

Pre-requisites of Extension Privatization

Privatization of government agricultural extension agencies is a major policy decision involving political, social and financial dimensions. The policy-makers will take such a decision with the expectation that the new arrangements will lead to more efficient and effective extension service than the one provided by the government. Therefore, before taking such decision, one must critically examine the presence or absence of the following pre-requisites. The results could strengthen or weaken the case for privatization of extension services.

Rise in the number of commercial farmers and cooperatives

It is important to determine whether the number of commercial farmers and large cooperatives, who are able and willing to pay for extension advice, has significantly increased in recent years. Private sector will be reluctant to invest in establishing advisory services for a small number of commercial farmers. Obviously, any private extension advisory service without a sizable market and demand for their services will have slim chances of survival.

Availability of worth-paying extension advice

It is common sense that farmers will not get much benefit from repetition of advice on routine farming practices which they have been following for years. They will be more interested in receiving recommendation and advice on new improved technologies, which they can try without much risk and increase their income, for which they will be inclined to pay. However, if it is business as usual in terms of the quality of extension advice, they will not pay for it and will prefer to continue the farming practices.

Complexity of farming practices

If the farming practices are so complex that they cannot be performed satisfactorily by farmers without the guidance of well trained and experienced extension advisors, who presently do not exist in the government extension service, then it will be a factor in favour of privatization. If farmers are engaged in routine farming, they will not really need frequent extension advice. But if they are engaged in a substantial post-harvest activity such as agro-processing enterprise, they will need specific technical guidance on operations and marketing of produce. Such guidance could come only from technical experts of private sector.

Presence of viable private extension service

The government extension agencies, if abolished for whatever reason, will require immediate replacement by a viable private extension advisory service or at least some transitional arrangements. Otherwise, farmers will be left without any extension support. Recently, when the functions of farm inputs supply were taken away from departments of agriculture in some countries while there was no alternate public or private mechanism for inputs supply, the farmers suffered immensely. It is equally important that the private companies and advisors have meaningful access to the sources of improved technologies, sufficient competent staff, and equipment and transportation facilities. Also, these service providers should possess knowledge of the latest trends of institutional and operational reforms in extension going on worldwide.

Farmers' capacity

If the farmers have to properly benefit from a privatized extension advisory service, they must have certain knowledge and skills necessary for dealing with the private advisors. These capabilities will include demanding particular extension services realistically, access to private companies and advisors, objective selection of quality advisors, negotiating fee, contract preparation, evaluating the extension advice received, resolving conflicts, up-to-date information on marketing, etc. As the experience shows in some developing countries, even when farmers formed their groups to strengthen their position vis-ḍ-vis private advisory services, they found themselves weak due to lack of capabilities such as mentioned above.

Government's readiness to assume new role

If a country goes from public to private in providing extension services to its farmers, it does not mean that the government has washed its hands off all responsibilities in extension matters. In fact, the responsibilities become greater although of different nature. The government's role under partial or fully private extension advisory services in a developing country will comprise policy advice, coordination, assurance of extension advice quality, outsourcing or contracting out (in case extension financing and extension delivery are treated as two separate functions), adequate supervision in the field, maintenance of a database of registered private advisory companies and individual advisers, formulation of rules regarding payment to extension service providers and conflict resolution between providers and recipients, environment protection, natural resources management, and possibly some others depending on the situation. It is imperative that the government machinery be in full gear to carry out these functions in order to protect mostly illiterate farmers from possible exploitation by the private sector, before any steps are taken towards privatizing extension services.

Costs of Unfeasible Extension Privatization

Unfeasible extension privatization means the privatization of public extension agencies in spite of the fact that all or most of the pre-requisites discussed in the preceding section do not exist in the particular country. Such an action will have serious social and economic consequences, some of which are as follows:

a) The farmers will lose the extension support irrespective of its quality that they have been receiving from the government agencies. They will have no convenient access to improved technologies and no solutions to their production problems thus causing a possible decline in the yields, which will deepen food insecurity and rural poverty.

b) The government extension agencies would most probably have been destroyed or seriously eroded in the process of privatization, making a large number of extension workers unemployed. Even if there is a newly created private extension sector, being still in infancy, it will not be able to absorb the entire or most of laid off government extension workers.

c) The private extension advisory services will not find sufficient number of client farmers who will be able and willing to pay for extension advice. As a result, the survival of private extension bodies will be at stake.

d) The government would not be able to protect farmers from the exploitation by the opportunist private advisors, creating conflicts in rural areas.

e) In the end, there will be neither public nor private extension service in the country, leaving the frustrated farmers at the mercy of nature. This could lead to social unrest.

f) The country will eventually suffer both economically and socially. There will be no one left in the rural areas to prepare farmers to meet challenges of globalization and market liberalization even though these developments are bound to affect the rural households.

g) The gap between rich and poor farmers will enlarge.

h) The revival of extension support will be a costly effort in every way.

Justification for Partial Extension Privatization

Why partial privatization?

If key pre-requisites for privatizing extension are met in some Asian countries with relatively well developed agriculture, a case may be made for partial privatization of extension. Total privatization will not be feasible due to the fact that millions of small farmers who are not able to pay for extension advice, apart from a lack of worth paying routine advice, still need public extension support. In addition, preferably, a lean government extension agency will be required not only to supervise and backstop the private mode of extension advisory services but also to cover common good practices such as environment protection, regulating and managing the use of natural resources, and biodiversity conservation which are of little interest to private service providers and to most of the farmers.

Visible trend towards commercialization of farming

Many Asian nations have achieved remarkable success in developing its agriculture sector while other remain relatively traditional and underdeveloped. In industrially more advanced Asian countries like Japan, South Korea and China, the commercialization of agriculture is more prominent than in countries like India which are also making progress in this area but at relatively slow rate. The trend to run farms as business and to start different types of agri-businesses is distinctly on rise in countries like Malaysia, Indonesia, Philippines, India and Pakistan. In Pakistan, for example, groups of small farmers have purchased pieces of land to have "agro-mall" built under franchise scheme. The agro-mall will serve as one-stop facility for farmers to buy fertilizers and other farm inputs, rent farm machinery, buy diesel, and even obtain extension advice. Malaysia and Indonesia are already engaged in the cultivation of major cash crops and fruits. Some of the countries have quite large and active cooperatives. The successful case of the commercial production of *salak pondoh*, a tropical fruit, and its profitable marketing in Sleman District of Yogyakarta Province of Indonesia by a group of farmers is well known.

Creation of a pool of private extension advisors

While partially privatizing extension for the first time, one of the main challenges is to create a pool of private extension advisory services providers without destroying or seriously undermining the public extension organization, apart from required organizational adjustments. As mentioned earlier, a number of non-public actors have now entered in the area of extension services delivery. They include well established commercial companies whose main business is not the provision of extension support but the sale of agricultural inputs and perform extension function within that context to promote their products. Such companies and NGOs can provide extension advice through contracts and outsourcing.

Another so far untapped source of private extension advisors is government retirees from various rural and agricultural disciplines like extension, rural development, community development, and general agricultural development. They can be given refresher training courses in their fields of experience, basic extension techniques and mechanism of privatized extension. The pool can provide private extension services on contracts with farmers' associations and cooperatives.

Improving Extension without Privatization

While discussing privatization of extension as a possible solution to the problem of unsatisfactory public extension agencies, the subject may be examined from a different perspective. Can the government extension agencies be improved without privatization with an aim to deliver what one expects from privatized extension? There is no doubt that the privatization alone is not going to solve all the present problems of government extension agencies.

The purposes of extension privatization are to alleviate financial burden off the government, make extension services more efficient and useful for farmers, to encourage private investment in national development as a matter of policy, and make the private service providers accountable if the recipients of the services are not satisfied. The success of privatization, especially for organizations like extension, which focus on bringing positive behavioural change among mostly illiterate farming communities, cannot be fully guaranteed. For example, the privatization of extension in several Latin American countries, after abolishing the public extension organizations, has not brought all the expected good results, to the frustration of farmers and policy-makers alike.

The meaningful improvement of a support service like agricultural extension, which has suffered for decades from low status, low salaries and benefits, inadequate operational budget, negligible opportunities for career development and poor facilities in the field, will require bold policy decisions and considerable investment in money, effort and time. Some of the measures for improving the existing public extension services will include the following:

a) Reform of the institutional and operational aspects of the national extension service in line with the latest global developments, challenges and opportunities.

b) Encouraging the participation of public, private and civil society institutions including NGOs in the provision of extension services through a pluralistic extension pattern.

c) Treating extension financing and extension advice delivery as two separate functions.

d) Downsizing of the extension organization if it is too large a bureaucracy.

e) Revision of the organizational structure of extension on the basis of subsistence and commercial farmers' needs and demands for extension advice and realistic determination of the types of staff needed.

f) Revision of the terms of reference of the entire staff.

g) In-service training of staff to bring them up-to-date regarding their new responsibilities under the reformed organization.

h) Possible organizational devolution or decentralization to an appropriate level.

i) De-concentration of staff from national and provincial levels and relocation of subject-matter specialists at organizational levels where they could be of maximum benefit to subsistence and/or commercial farmers.

j) Optimum application of modern information technology in support of field extension activities and institutional linkages.

k) Use of participatory, gender-sensitive group extension approaches including bottom-up extension programme planning at grassroots level by farmers' groups with facilitation by extension workers.

l) Provision of essential facilities including adequate operational budget and transportation to field extension staff.

m) Development and use of original, location-specific, gender-sensitive, participatory and low-cost extension methodologies and materials.

n) Strengthening of operational linkages between extension and research, marketing and other relevant institutions.

o) Improved career development opportunities and other incentives for extension professionals.

p) Objective accountability of extension workers to the farmers.

q) Improvement in the pre-service education in extension at academic institutions to bring its curricula, instructional methods and materials in line with the latest developments in extension.

Conclusion

Unsatisfactory performance of most public extension services and the pressure from some donors has triggered the thinking of policy-makers towards possible privatization of government extension agencies. Privatization demands major policy decisions, with short- and long-term political, economic and social implications. In addition, unless certain pre-requisites are met, privatization may cause more damage than good. Most Asian countries have a huge population of subsistence farmers. Their transformation into business-oriented producers will justify privatizing extension, but the transformation itself will remain a colossal challenge for quite some time to come.

Under prevailing conditions, partial privatization of extension services may be justified for commercial farmers and large cooperatives in spite of their relatively small number, because the traditional public extension organizations cannot satisfy their more advanced extension and training needs. However, even under partial privatization, the present role of government will change both for protecting the interests of farmers and taking responsibility for common goods such as natural resources regulation and management, environment protection and biodiversity conservation.

It must be kept in mind that the privatization alone is not the solution to all the problems that the government extension agencies suffer from. Full-fledged institutional and operational reforms are required to enable the national extension service, irrespective of its being public or private, singular or pluralistic, to serve the challenging learning needs of farmers emerging from the latest international developments like globalization, market liberalization, decentralization and participation.

References

Dinar, A. and G. Keynan (2001). Economics of paid extension: Lessons from experience in Nicaragua. *American Journal of Agricultural Economics* 83: 769-776.

Dorman, E., K. Amezah, and J. Hesse (2003). *Contracting out extension in Ghana: The first steps.* Paper presented at the 25th International Conference of Agricultural Economists, Durban, South Africa; August 19.

Eaton, C. and W. Shepherd. (2001). Contract farming: Partnerships for growth. *FAO Agricultural Services Bulletin No. 145*. Rome.

Neuchatel Initiative Group (2002). *A common framework for financing agricultural and rural extension*. Neuchatel Initiative, Uppsala, Sweden.

Qamar, M.K. (2000). Agricultural extension at the turn of the millennium: trends and challenges. In *Human Resources in Agricultural and Rural Development*. FAO, Rome.

Qamar, M.K. (2002). *Global trends in agricultural extension: Challenges facing Asia and the Pacific Region.* Keynote paper presented at the FAO Regional Expert Consultation on Agricultural Extension, Research-Extension-Farmer Interface and Technology Transfer, held in Bangkok; 16-19 July.

Qamar, M.K. (2005). *Modernizing National Agricultural Extension Systems: A Practical Guide for Policy-Makers of Developing Countries.* FAO, Rome.

Rivera, W.M. (2001). The invisible frontier: the current limits of decentralization and privatization in the developing countries. In F. Brewer (Ed.), *Agricultural Extension: An International Perspective.* North Chelmsford, MA, USA, Erudition Press.

Rivera, W.M. and J.W. Cary (1997). Privatising agricultural extension. In B. Swanson ed. *Improving Agricultural Extension: A Reference Manual*. FAO, Rome.

Rivera, W.M., M.K. Qamar and L.V. Crowder. (2001). *Agricultural and Rural Extension Worldwide: Options for Institutional Reform in the Developing Countries*. FAO, Rome.

Rivera, W.M. and W. Zijp (2002). *Contracting for Agricultural Extension: International Case Studies and Emerging Practices*. Wallingford, UK; CAB International.

Van den Ban, A.W. (2002). Different Ways of Financing Agricultural Extension. *ODI, AgREN network paper 106b*. London.

Part 3

Supporting Poor farm families

11

Agricultural Extension and the Underprivileged Farmers: A Case for Change in Extension Paradigm

D.V. Rangnekar

Introduction

Agriculture extension started in India soon after independence, in support of a variety of development activities and the focus was mostly singular viz. transfer of technology for improving food production. A good example is the 'Training and Visit' (T and V) system of agriculture extension, which was introduced in early eightees to support increase in food production by popularising 'adoption of hybrid seed-fertiliser technology'. The type of extension programmes taken up in those early years are also referred as 'first generation extension' and characterised by singular focus, uniform recommendations and top down approach. There was realisation that while the impact of T&V was impressive in irrigated areas it was not discernible in rainfed areas where farming systems are more complex and majority farmers are resource poor and the linear model of technology transfer did not work well. Subsequently the need to take into consideration other factors like agro-ecology, farming situation etc., while making recommendations, was felt and thus emerged the 'second generation extension'. The T and V system is still a subject of debates and discussions and many have analysed the system in detail (Hayward, 1989, Nagel, 1996.). While many suggestions were made by the analysts to change the approach of agriculture extension, the major thrust continues to be on transfer of technologies and recommendations (also referred as Lab to Land approach) without much consideration of varying situations in agriculture. It was presumed that answers/technologies for all types of situations/ problems are available off the shelf, ready to be transferred to farmers. Another presumption still prevalent is that answers/ solutions to the problems in agriculture and the knowledge needed for this purpose is only with researchers and laboratories. There was very little participation of the farmers in the extension process and their experience and wisdom was hardly taken into account. It soon became evident that adoption of technologies and recommendations by the small farmers has been very limited. There was some introspection resulting in recognition of farmer's knowledge/wisdom/innovations and need to encourage their involvement was strongly recommended (Chambers, 1993; Cornwall et al., 1994). Subsequently the need to take into consideration other factors like agro-ecology, farming situation, etc. while making recommendations was felt and thus emerged the 'second generation extension'. Since last decade or so attempts have been made to develop extension systems with farmer participatory and systems approach through some projects in a few pockets of the country. And thus 'the Third generation extension' is emerging. The systems approach also drew attention to the new challenges posed by 'changing farming situation due to changes in market demand for products and also due to Agreement on Agriculture signed by India under WTO'. While many extension specialists and analysts have also pointed out constraints (like availability of funds) faced by the extension systems. However, very few reports discuss in detail the ways to extend benefits to the resource poor farmers and meet the challenge of poverty reduction.

This paper is not meant to be an erudition or a critique on agriculture extension, but ' an honest expression of concern regarding neglect of underprivileged' and an attempt to present 'a case for change in extension paradigm, so that agriculture can provide a pathway for sustainable livelihood development for the underprivileged families'. The paper is based on limited literature search and more on observations and experience in the field, gained through long standing involvement in agriculture based rural development programmes that provided opportunities for interaction with the underprivileged families.

The reasons for concern that provoked me to write this article are - that

- a review of literature on agriculture extension shows very few papers have assessed extent of benefits of extension reaching the underprivileged and even these make only passing reference to this issue (I admit that I may have missed some publications on this aspect);
- the field reality shows 'neglect of the underprivileged and concern about who is benefiting', although the underprivileged represent

sizeable portion of the rural population (about 30%) and their contribution to agriculture produce is substantial (food crops);

- most papers discussing changes in agriculture extension systems have mainly dealt with the need to meet challenges posed by WTO and varying and changing farming situations;
- I observe that 'most extension officers function as postmen – delivering what is given to them by experts'; 'Transfer of available technologies and recommendations' still continues to be the main purpose and 'most recommendations are uniform irrespective of variability in farming situation';
- reports by economists (Mellor, 2005 and others) indicate that in India whole farm development is the best option for sustainable development of the livelihood of the underprivileged families, for whom alternate employment opportunities are very limited;
- steady increase in demand for livestock products is reported and livestock is the mainstay for majority of the underprivileged; thus they can benefit from increase in demand of livestock products;
- livestock extension is grossly neglected and needs to be strengthened to help the underprivileged.

To express my concern explicitly I propose to discuss a few randomly selected papers (published in the last decade or so) that analyse different aspects of agriculture extension; since these papers reflect views and concerns of the experts on extension. In the end I propose to refer a few papers/reports that show how alternate/appropriate approaches enabled underprivileged to benefit from extension services.

The Underprivileged and their farming systems

The Underprivileged word, used in the paper, refers to all those rural families that belong to socio-economically weaker strata.

There are very few studies reported in literature on farming systems (crop-livestock production) of the underprivileged rural families. Observations show that farming systems of the underprivileged are somewhat different from those of resource rich farmers and main characteristics of the farming systems of the rural underprivileged are presented in box – 1.

Box - 1

Features of farming systems of rural under-privileged families

1. Mixed farming system with diversified crop and livestock activities.
2. Assured subsistence is preferred over higher productivity, which is considered risky.
3. Low external input — low output system (highly internalised), maximum use of available resources like crop residues, feed, labour, manure, draft power etc.
4. Crop and livestock production has multipurpose objectives. Crops to produce grains for humans and straw for animals, while livestock to provide nutritious food, draft power, fuel and fertilizer.
5. Crop and livestock that niche well with local conditions are chosen to minimize risks (based on generations of experience).
6. Traditional systems of crop and livestock management are preferred and adoption of new technologies is very low due lack of confidence.

7. Livestock are important part of the system and though their output is low but it represents major share of daily cash income to family.
8. Women play a major role in farming and particularly of livestock production and sale of produce.

Adopted from Rangnekar (1998)

These systems aim at optimising use of limited resources (material and labour) available, minimise external inputs and avert risks; as against maximisation of profit by the resource rich. 'Diversification and Internalisation' and 'low adoption of new technologies and scientific recommendations' can be said to be other features of the farming systems. Findings of a study carried out in Rajasthan by Singh and Varghese (2004) provide a good picture of existence of diversification in agriculture in resource poor area. They report that majority of the farmers had diversified crop (3 to 6) and livestock (2 to 3) activities, except in desert and tribal dominated districts. According to them diversification is a way of risk coverage and maximising utilisation of resources, In such situations extension programme/ officers with commodity approach and trying to transfer/ disseminate singular recommendations would neither be useful nor effective. There are very few studies on reasons for low adoption by the underprivileged and observations indicate that non- relevance, risk perception and lack of confidence about benefit from the technologies are some of the main reasons. Moreover, most of the technologies and recommendations are based on research conducted in institutions where conditions are very different from that of the resource poor farmers.

The underprivileged families are 'classical examples of state of deprivation'. They suffer from deprivation of self- confidence, opportunities to acquire skills, information, resources, organisational support for services, credit as also of proper status in the society. Those who have worked in rural areas would be aware that social aspects have a strong bearing on production systems and particularly of livestock as they have stronger social linkage. Hence, development programmes should plan for developing all the five assets (natural, social, human, physical and financial capitals) related to livelihood, as described by DFID (Carney,1998), rather than confining to improving family income. Integrated agriculture development is able to achieve such results, with support of proper extension effort, as is shown by NGOs like BAIF (Hegde,2005) and some cooperatives (there would be similar other projects that I am not aware of).

Definitions of Agriculture Extension

Agriculture extension has been differently defined and the definition varied from time to time, to the extent that it is difficult to find a straightforward definition. Major goals and stage of agriculture development as also the perceptions of extension specialists influenced the way it was defined. The definitions of extension ranged from 'persuasive transfer of technology to human development'. For example Nagel (1995) described extension as 'the organised exchange of information and the purposive transfer of skills', while Garforth and Lawrence (1997) provide a much broader definition. They state that ' agriculture extension is the conscious provision of information and communication support to rural users of renewable natural resources' and 'involves offering advice, helping farmers analyse problems and identify opportunities, sharing information, supporting development of farmer groups and facilitating collective action'. Van de Ban and Hawkins (1998) state that 'it should be helping farmer in making better decisions in changing systems or improving them'. Holistic concept of agriculture development creeped into the thinking of extension specialists during this period and aspects like 'Agriculture Information Systems' and ' Innovation Systems Framework' were incorporated.

Analysis of agriculture extension approaches and suggestions to improve it

Agriculture extension has been analysed by a number of extension specialists, economists and sociologists in relation to its approach, impact on production, adoption of technologies and scientific recommendations, economic benefits, awareness etc. and lately for challenges it faces due to Agreement on Agriculture under WTO. Recommendations are made for changes in extension programmes and these vary with time and perceptions of persons and influenced to a large extent by the issues in vogue and by major challenges faced by agriculture. Papers indicate realisation that farmer needs now go beyond information and technologies related to agriculture and extension is expected to help/facilitate organisation development, access to market and technical information etc. These papers also point out limitations of extension systems like trained manpower and funds and other factors influencing its working.

he FAO publications – There are some useful publications of the FAO based on compilation of papers that analyse agriculture extension and provide recommendations to make it more effective (there may be some more that I may have missed). One of the publications based on expert consultation on extension (FAO, 1990) compiles 6 papers dealing with a variety of subjects ranging from evolution and current status of agriculture extension, in depth case studies of different approaches to extension to experiences of donor agencies like World Bank, USAID and the FAO. The World Bank paper discusses in detail performance and limitations of T and V, besides other projects. Most of the papers discuss changes needed to make agriculture extension more effective and have pointed out constraints/limitations faced by the extensions systems. The report recognises that economic pressure and stress on showing results has forced extension to focus on technology transfer, increasing agricultural production and on commercial farm sector. Its role in human resource development and helping the resource poor is getting neglected. However, the report does express concern that some donors give emphasis to technology transfer rather than recommending a balanced approach.

The third edition of the Reference Manual on Agricultural Extension published by the FAO (Swanson et al., 1996) is a comprehensive publication with 23 chapters written by 38 authors, covering most of the aspects related to extension. The publication provides an overview of extension, economic contribution of extension to agricultural and rural development and discusses improvement in extension management. There is a chapter emphasising adoption of 'needs assessment process' and developing extension programme to meet identified needs. There are chapters discussing 'alternate approaches to organising extension' and one describing 'emerging and important role of nongovernmental organisations' in the extension process and in organising small and marginal farmers.

While both the reports recommend that extension should pay equal attention to the resource poor and women farmers, none of the papers discuss in detail extent of benefits received by the underprivileged and approach or changes needed to ensure that they derive benefit. The recommendation in the reference manual (Swanson et al., 1996) is more related to variations in farming situations rather than socio-economic condition that puts the underprivileged in a disadvantageous situation.

Publications from the ICAR organisations and scientists: A few randomly chosen papers are discussed below and these presumably reflect thinking that prevails in the country regarding agricultural extension and need for changes in extension. One of the policy briefs (no. 9) by Sulaiman and van den Ban (2000), published from National Centre for Agricultural Economics and Policy Research (NCAP) describes that agriculture in India is going through some major changes due to shrinking resource base, changes in food consumption pattern, liberalisation of agricultural trade and changing farming system. Consequently farmer's needs for support from agricultural extension are changing and to remain relevant and effective the extension departments of various states have to bring in some major reforms. The extension staff has to acquire knowledge on aspects like market demand, policies and prices for various commodities as also on agriculture related policies besides production technologies. They also need to acquire social skills required to undertake needs assessment, conflict resolution, group formation, participatory skills etc. It is stated that 'Public sector extension still has a major role to play' while privatisation of extension needs to be considered. Another policy brief (no. 17) by Sulaiman and Hall (2004) discusses their concept of 'Extension Plus' so that a diverse set of objectives of extension can be fulfilled. In this paper they make two important observations; one related to need to realise that a single model of extension cannot work for all kinds of situations and new approaches have to emerge locally; the second point relates to the need to look into poverty reduction and enhancing voice of the poor. Sulaiman and Hall (2004) describe a range of extension initiatives from the public and private sectors 'that display the expanded agenda embodied in the concept of extension plus'. They have briefly described initiatives taken by the Government, agribusiness, Financial and NGO bodies as 'Extension plus example from the field'. They emphasise need for 'diversity of approaches and arrangements' and have illustrated the key shifts required need for new culture, while voicing concern about limitations of public sector.

The underprivileged and Livestock production

A sizeable population of underprivileged keep and depend on livestock and agriculture extension has generally neglected livestock production (Rangnekar 1995). Demand for livestock products in developing countries like India is increasing with improvement in economy and changing food habits and there is scope for resource poor to improve their livelihood by taking benefit of this development (Delgado etal, 1999). However, according to Thomas and Rangnekar (2004) the underprivileged need to overcome some constraints (economic, technological and social) before they can take benefit of increasing demand for livestock products. While citing some examples of attempts being made to overcome these constraints they point out need to strengthen livestock extension to enable the underprivileged to improve productivity and quality of livestock products.

Parthasarthy Rao etal (2004) constructed a crop livestock system typology to better understand small scale mixed farming systems in selected South Asian countries (Nepal, India and Sri Lanka). They find that the system is in different stages of evolution in view of increasing demand for high value products like fruits, vegetables and livestock products, providing opportunity for smallholder farmers to improve their income. They also find that agro-climatic, technological and socio-economic factors drive these changes and productivity of livestock will have to be improved to take full benefit. They indicate that low productivity of livestock with smallholder farmers is reflection of non-adoption of available technologies or their uptake has not been sustainable, probably because they were introduced without proper understanding of variation and complexity of the systems.

Rangnekar (2003) while discussing buffalo production from a sustainable livelihood perspective have drawn attention to the fact that buffalo is one of the most preferred of livestock and can play a crucial role in improving livelihood of the underprivileged. However, there is need to change research-extension paradigm to enable them to take desired benefit.

Morton et al., (1997) studied livestock production extension in Kenya, India and Burkina Faso with the objectives of reviewing and assessing the systems and evaluated impact. They conclude that in most of these countries although some national livestock extension services are already in place but they are mostly serving the wealthier farmers and the resource poor are deprived of their benefit. They report that production related extension is rather poor in most cases. They recommend that 'donors should seek to enhance integration of crop-livestock systems through restructuring of the systems, retraining of staff and strengthen research-extension linkages'. Ravikumar

(2005) has reviewed livestock extension in India as part of his thesis work and studied existing system in detail in Tamilnadu. He points out that livestock extension is grossly neglected and several constraints like lack of funding, trained manpower and orientation of the staff. He reports that services related to health aspects are fairly well organised and most of the funds are used for these services, production related extension is neglected. He observes that much of the benefit of public sector services is taken by the resource rich as was also reported by Morton et al., (1997). It is obvious that considerable reforms are needed in the extension system to help the underprivileged.

Approaches that benefited resource poor farmers – *some examples*

Maarse et.al; (1994) documented their experience of changing from conventional transfer of technology approach to participatory and systems approach of extension, in a dairy development project in Kenya. The document shows that to transfer technologies in a mechanical manner was not effective and the adoption was low. Studying production systems, involving farmers and scientists together to find ways of improving productivity with small farmers has been effective. An interesting outcome is perceptible change in the attitude of scientists as well as farmers towards each other, resulting in critical study of constraints and development of appropriate solutions.

The BAIF, an NGO, extensively involved in agriculture based rural development work in many states, has reported success with an approach of developing a team of para-extension workers (PEWs)/ link farmers (LFs). Much of BAIF's involvement is with underprivileged families from rain-fed areas. These

PEWs/LFs are chosen by the community and trained by the BAIF to study the farming situation, identify constraints and participate in deciding interventions and extension messages according to the needs and in consultation with technical persons. These PEWs/LFs are also involved in disseminating solutions/ technologies chosen for overcoming constraints and improving productivity (Hegde, 2004). Results of these programmes show that the organisations and extension teams sensitised to adopt pro-poor, need based and participatory approach produced desired results. The extension team also provided support to develop post-production activities, market linkages and in securing credit facilities in addition to working on improving farm productivity. Another crucial factor contributing to the success is the credibility earned and rapport established by the team with the rural families. Sulaiman and Hall (2004) have also cited BAIF's experience and referred it in his policy paper advocating development of 'Extension plus approach'.

In the state of Andhra Pradesh the NDDB facilitated adoption of 'Farmer Participatory and Need based Extension' (FPNE) approach in four dairy producer's cooperative unions involving dairy supervisors, who are also dairy producers. The objective of this effort is to improve quality and productivity of milk by providing services and information to the producers according to the needs identified by them. It was also envisaged that such efforts improve effectiveness and efficiency of services and sensitise the Dairy unions to closely involve producers and particularly the women in the process. The FPNE is a process that starts with 'analysis of situation' as it exists in an area, enabling full understanding of the existing situation and is followed by identification of constraints. Roy and Rangnekar (2006) while reporting details of the approach state that the process has sensitised the dairy unions to take differential and need based approach while dealing with resource poor producers and promoted closer interaction between the staff and the producers.

In conclusion it can be said that there is a case for change in development-research and extension paradigm through strategic planning, considering variability in farming and livelihood systems of the underprivileged. It is only through a strategy which is farmer centred, pro-poor, sensitive to socio-cultural and gender aspects and is need based that benefit can be extended to the underprivileged. However, this is easily said than done and needs sustained effort at changing mindset and all have to be sensitised to the needs of the underprivileged.

References

Carney, D. (1998). Implementing the sustainable rural livelihood approach, in Sustainable Rural Livelihoods. Papers presented at the Department of International Development's Natural Resources Advisor's Conference, July, 1998. , London, D.F.I.D., pp. 3-26.

Chambers, R. (1993). *Challenging the professions, frontiers for rural development*. Intermediate Technology Publications, London.

Cornwall, A.; I. Guijt and A. Wellburn. (1994). Acknowledging process: methodological changes for agricultural research and extension. In Scoones, I. And J. Thompson (Eds.) *Beyond Farmers first, rural people's knowledge, agricultural research and extension practice.* Intermediate Technology Publications, London.

Delgado, C., Rosengrant, M, Steinfield, H., Ehui, S. and Courbois, C. (1999). Livestock in 2020, the next food revolution. Dicussion paper 28, I.F.P.R.I., Washington, D.C.

F.A.O. (1990). Report of the global consultation on Agricultural Extension, held at Rome, 4 to 8 Dec. 1989. Food and Agriculture Organisation of the United Nations, Rome, Italy.

Garforth, C and A. Lawrence (1007). Supporting sustainable agriculture through extension in Asia. *Natural Resources perspectives*, No. 21, June 1997.

Hayward, J.A., (1990). Agricultural Extension. The World Bank's Experience and Approaches, in *Report on Global Consultation on Agricultural Extension*. F.A.O., Rome, pp. 115-134.

Hegde, N.G. (2005). Innovative approaches for poverty alleviation. BAIF Journal (26), April, 2005, pp. 1-2.

Maarse, L. , W. Wentholt and A. Chibudu. (1998). *Making change strategies work.* Royal Tropical Institute, Amsterdam, The Netherlands.

Mellor, J.W. (2004). Agricultural growth and poverty reduction: the rapidly increasing role of small holder livestock. In *Livestock and livelihoods. Challenges and oppportunities for Asia in the emerging market environment*. Ed. V. Ahuja. Pub. N.D.D.B., Inidia and F.A.O., Rome pp. 30-50.

Morton, J., R. Mathewman and D. Barton (1997). Livestock production extension: issues, case studies and policy options. NRI Socio-economic Series 12. National Resources Institute, Chatham, UK.

Nagel, U.J. (1996). Alternative approaches to organising extension. In Swanson, B.E., R.P.Bentz and A.J.Sofranko, Eds. *Improving Agricultural Extension: A Refrence Manual*, F.A.O., Rome.

Parthasarthy Rao,P., P.S.Birthal, D. Kar, S.H.G.Wikramratne and H.R.Shreshta (1999). Increasing livestock productivity in mixed crop-livestock systems in South Asia. Report of a project. National Centre for Agricultural Economics and Policy Research, New Delhi, India and International Crops Research Institute for the Semi-Arid Tropics, Patancheru, India.

Rangnekar, S.D. (1998). Women in agriculture and modern communication technology. Proc. of a workshop organised by The Danish Agricultural Forum at Tune Landboskole, Denmark, 1998. pp. 26-37.

Rangnekar, D.V. (2003). Buffalo production from sustainable livelihood perspective. Proc. 4th Asian Buffalo Congress: Lead papers. Pp. 254-259.

Ravikumar, R.K. (2005). Livestock extension activities under the state departments of animal husbandry in India: an institutional analysis. Ph.D. thesis submitted to the Indian Veterinary Research Institute, Izatnagar, UP.

Roy, S. and DV Rangnekar (2006). Farmer Participatory Need based Extension (FPNE) Approach: A Sustainable Model Adopted by Cooperative Milk Unions in Andhra Pradesh, India (paper submitted for publication).

Singh, R.P. and Varghese, K.A (2004). Diversification of economic activities and farm household income in Rajasthan. Socio-economic and agricultural policy planning cell, Rajasthan College of Agriculture, M.P.University of Agriculture and Technology, Udaipur.

Swanson, B. E; R.P. Bentz and A. J. Sofranko (1998). Agricultural Extension: a reference manual. *Food and Agriculture Organisation of the United Nations, Rome.*

Sulaiman, R.V. and A.W.van den Ban (2000). Agricultural extension in India – the next step. *Policy Brief no. 9. National Centre for Agricultural Economics of the ICAR, New Delhi.*

Sulaiman, R.V. and A. Hall (2004). Towards Extension plus: opportunities and challenges. *Policy Brief no. 17. National Centre for Agricultural Economics of the ICAR, New Delhi.*

Thomas, D. and Rangnekar, D.V. (2004). *Responding to the increasing global demand for animal products: implications for the livelihoods of the livestock producers in developing countries. In* Responding to the livestock revolution: the role of globalisation and implications for poverty alleviation. *(Ed.) Owen, E. et al., British Society of Animal Science, publication no. 33.*

Van den Ban, A.W. and H.S. Hawkins (1998). Agricultural Extension. *Oxford : Blackwell Science and CBS Publishers and Distributors, New Delhi.*

12

Thailand Research Fund: A Research Funding Agency and Its Influence on Agricultural Extension Projects

Avorn Opatpatanakit

Situation of Agricultural Community in Thailand

Having grown up in farmer family in Northern Region, I have observed for a long time the situation of the agricultural community in Thailand. Similar to many other (post) modern societies throughout the world, Thai society, rural areas in particular, has been confronted with problems affected by rapid and complex changes under the mainstream modernization policy during last fifty years. It is widely observed that there are more evidences of such economic crises as indebtedness, landlessness, bankruptcy, and poverty. Similarly, such social crises as loss of indigenous knowledge, unemployment and urban migration, increased sexual trade, increased drugs abuse and AIDS victims could be easily found in both rural and urban areas. While deterioration of natural resources and environment, scarcity of water resources, and soil depletion are concrete evidences of the crises of natural resource management (Opatpatanakit, 2001).

If one asks the elder farmers in the village or sub-district communities in Chiang Mai and Lamphun province, about "what do they worry about?". The answers always are as follow:

- high production cost of dairy cows, especially cost of concentrate and roughage, but the price for fresh milk is very low,
- high cost of *longan* production but low price gained by farmers, yet the middlemen and the exporter become rapidly rich within a year,
- loss of indigenous variety of rice, or cow such as "Kokhao Lamphun" breed,
- less fish and weeds in channels where concrete check-dams have been built replacing the bamboo ones,
- less forest and long-period of drought or flooding,
- scarce food from the forest nearby the village,
- fighting among cousins for water for rice and *longan* fields
- lost of such cultural farm-oriented activity as reciprocal labour (called *ao wun* in the North region), in agricultural production and other family as well as village affairs,
- less youth will continue to take care of their family farms because most of them prefer to work in the nearby factory.

Mainstream Agricultural Extension as Threat

Such phenomena of farmer's worry reflect the failure of the agricultural development system and agricultural extension in northern region of Thailand, particularly in Chiang Mai and Lamphun province. Most conventional agricultural extension projects appeared to be of a top-down nature by which agenda, areas and budget of major agricultural development and extension projects were mostly set by the policy makers at the ministry and department levels. Agricultural extension workers at the provincial, district, sub-district levels would then translate this policy into practices through conventional training designed by extension workers, involving farmers as passive participants.

The important factor contributing to the lack of success of such agricultural extension projects is that farmers have not had a chance to participate in the decision making process of such learning. Mainstream agricultural extension projects provided little room for the farmers to share their experience and local wisdom, no matter how well they understand their own problems and have developed their local wisdom throughout their life. Such top-down extension projects were normally undertaken in all communities throughout the province without identifying community capital and needs, and analyzing the situation of farmers under the rapid changing world. Involved farmers did not learn much from such conventional training process. They do not even think about *"why they should attend the training?, what will they learn?"* This is because they were paid to participate. Accordingly, they had lost their self-confidence. Often they cannot initiate to think of even such a simple task like critical analysis on what are the root causes of the problems.

Being passive participants, it is less likely for farmers to develop the interactive learning process by which their capacity and skills in intelligently creating improvements could be increased. Clearly, for farmers, these agricultural extension projects were less useful. Moreover, under the *'top-down' bureaucratic* and *centralized administration*, agricultural extension projects have to finish by the end of the fiscal year. This limited administration has failed to deal with the fact that the problems of farmers were far more complex, diverse and unpredictable and hence required an even more flexible facilitating process (Opatpatanakit, 1999; Opatpatanakit, 2001). Thus, these agricultural extension projects have contributed mainly to the extension scheme itself and benefits external extension workers rather than farmers.

Application of Community-Based Research (CBR) in Agricultural Extension Project

The community-based research management has been employed for six years by the Thailand Research Fund (TRF) which has been established in response to the 1992 Research Development Endowment Act as a juristic body outside the government administrative bureaucracy. Its autonomy provides great flexibility to develop research management approach that effectively supports user-oriented research, particularly community-based research management scheme at the local level. Contrary to conventional research management systems employed by other research funding agencies, community-based research management (CBRM) scheme emphasizes the support of the CBR project under *three main conditions* :

- thematic concern and research question are identified by stakeholders of which the locals constitutes an integral part,
- local people are active co-researchers and decision makers of the CBR project, and

- the major portion of CBR project is a series of diverse actions that innovatively develop alternative approaches to solve the stakeholders' problems.

The outstanding mechanism of TRF in CBR management is its coordinator who effectively facilitate stakeholders to empower themselves to develop learning process, knowledge and community organization, which are crucial components of local community sustainability. The participatory action research is recommended as a research methodology by which the locals could be facilitated to become the decision-maker of a continuous cycle of planning, action, observation, and critical reflection and re-planning. Becoming the coordinator of the Thailand Research Fund (TRF) in 1996, I have applied the Community-Based Research in agricultural extension projects in various communities in Chiang Mai and Lamphun province. There are three main stages of CBR management.

Up-stream management: The empowering process normally begins with the development of the CBR project proposal, which usually takes from four to twelve months. The TRF coordinator, who becomes the facilitator, would design her/his facilitation as the learning process by which stakeholders are enabled to analyze the problems and thematic concerns of each group/community through such activities as dialogues, focus group discussion, and informal meeting and workshops. Various concerned people and organizations are encouraged to provide important data and information that help to identify the group's problems, their root causes and then only concerned themes are identified. The core research problems would then be formulated and the CBR project would then be granted by TRF.

Mid-stream management: During the implementation process, the core researchers have practiced collected data and information in accordance to minor and sub-research questions by documenting the process and output of each activity. Such process is undertaken through the observation of problem-solving activities, critical reflection towards each activities process and output through dialogues, small group discussions, informal meetings and workshops. Critical reflection is organized after each activity as to reflect on the process and output especially on what needs to be adjusted or improved in the next activity. The TRF coordinator has facilitated such critical reflection during the first two or three months, and then the core researchers would do it by themselves. Consequently, the core researchers have analyzed data and information in order to answer research questions. With the support of the TRF coordinator and external peer, they also synthesized the conditions contributed the empowering process.

Down-stream management: The TRF coordinator has facilitated the dissemination of CBR project's experiences by various means as articles for newspaper and journal, pocket book, dialogue through community radio, discussion in the meeting, workshop, annual conference, being a learning group or community by which other people can visit and exchange lesson learned.

The Influence of TRF towards Agricultural Extension Project through the CBR Project

The following paragraphs will illustrate the influence of TRF towards agricultural extension project through CBR projects which I have been facilitated in Lamphun and Chiang Mai province.

Kokao Lamphun Group research project

It is important to note that CBR approach emphasizes the "research question" as an effective tool that could help an agricultural extension project to start in the appropriate direction. For example, another community-based research project that has supported promoting the agricultural extension project to promote the local cattle breed called "Kokao Lamphun" back in Lamphun province. The core researchers of Kokao Lamphun Group research project, which had been supported by TRF during December 2003 to August 2004, composed of cattle farmers, the Livestock officer in Lamphun province, academics from Faculty of Agriculture, Chiang Mai University. They were facilitated to identify the major, minor and sub research questions.

The project research questions of Kokao Lamphun Group found to be "how Kokao Lamphun Group's members could effectively manage the group?" while the minor and sub research questions includes:

- *Group management*: How to support each member to perform well with his/her role? How to manage the group's fund, how to train the group member who is taking care of group's account and process documentation,
- *Cow raising*: What is the member's experience in raising cow?, How to raise cow on his/her farm? How to improve the member's capacity to observe the growth of cow and conditions that contribute the growth rate? How to convince the livestock researcher to collaborate in this CBR project?, How to encourage the district and provincial livestock officers to collaborate with the group's members?,

- *Roughage production*: What kind of grass should be planted? How much each cow consumes per day? How many tons of grass needs to be produced for the whole year? What should be the cheapest way to conserve nutritious grass for the dry season? What is the production cost of grass plantation?,
- *Learning of school students on Kokao Lamphun production*: How to facilitate the learning of school students? How to develop the so-called local curriculum about Kokao Lamphun production?,
- Can Kokao Lamphun production be an alternative way to decrease debt, or means to initiate agricultural tourism?.

These research questions had helped the core researchers to develop an more reasonable and effective action plan. At the moment, the Kokao Lamphun group has expanded their collaboration to the Lamphun Chief Executive Officer (CEO) and Pangola farmers.

Chaiprakaan Dairy and Agricultural Cooperatives Program

Similarly, dairy farmers in Chaiprakaan district, Chiang Mai province had started raising dairy cattle in 1995. Dairy milk production becomes the major activity of the Chaiprakaan Dairy and Agricultural Cooperatives by which 20 tonnes/day of fresh milk are produced by 185 farmers. However, raising dairy cattle is new to farmers, the first four years was a struggle period as there were many problems, for example (Opatpatanakit and Prason, 2003):

- infertility
- low milk production,
- less skill in estrous detecting,

- high incidence of mastitis,
- insufficient roughage in dry season,
- environmental problems as bad smell in rainy season; contamination of water in surface well which creates a conflict with the neighbors.

The core researchers of Chaiprakaan Dairy Cooperatives found to be the president, committee, staff, district Livestock officers, Cooperative district Promotion officers. They then identified research questions in a participatory working objective and expected output, and action plan, revised every six-months, for the following sub-projects:

1. Dairy health and productivity program I
2. Dairy health and productivity program II
3. Production of total mixed ration (TMR): corn silage
4. Farm accounting system relating to health and productivity program
5. Approach for technology transfer
6. Feasibility study on local marketing channel
7. Participation of local school in dairy production.

Having participated in the research projects, the dairy farmers' skills and knowledge in farm management had been improved. This contributed to a higher milk quality. Similarly, the marketing feasibility study helped the committee to decide to stop the construction of processing factory. Furthermore, the participation of local school in dairy production research project has facilitated schoolteachers and students to know and understand more about the situation of dairy production.

The district Livestock officers have developed dairy health and productivity program I and II with the staff of the Veterinary Faculty, Chiang Mai University which contributed to more effective farm management of the first 25 farms and to develop an appropriate new approach to technology transfer to other farmers. Similarly, the district Cooperatives Promotion officers also developed farm accounting system relating to health and productivity programs I and II with the academics from Department of Agricultural Economics, Faculty of Agriculture, Chiang Mai University. These district extension officers have integrated research process to fulfill the objective of the extension process.

Dairy farmers of 6 dairy cooperatives in Lamphun and Chiang Mai province encountered a shortage of roughage each year. It was even so serious in 2004 that the farmers had to buy Pangola grass from the Central region. The Livestock officers and the Cooperatives Promotion officers in Lamphun province, the committee of dairy cooperatives, academics from Faculty of Agriculture, Chiang Mai University had developed the research project for Pangola production in Lamphun and Chiang Mai province to serve dairy cooperatives. This research project has been supported by TRF. By the end of September 2005, there are 120 farmers who grow Pangola grass for their own cows and to sell to cattle farmers. Furthermore, these researchers also develop the Pangola marketing system.

Furthermore, the Livestock officer and the Cooperatives Promotion officer in Lamphun province and academics from Faculty of Agriculture, Chiang Mai University developed a research project on integrating Pangola grass with cattle raising in Longan farms as to be an alternative agricultural activity for Longan farmers due to its low price. This project is developed based on the lessons learnt by the Kokao Lamphun Group Community-Based Research project discussed above.

Feasibility study on integration of cattle raising with Pangola grass production in Pasak sub-district, Lamphun province

Additionally, according to the government policy on "One District One Bio-fertilizer Factory", with the support of TRF, the Livestock officer and the Cooperatives Promotion officer in Lamphun province develop the community research project on "Feasibility study on integration of cattle raising with Pangola grass production in *Pasak* sub-district, Lamphun province". This research project aims to find out how to produce cattle dung for the Bio-fertilizer Factory which will be used to encourage rice farmers in *Pasak* sub-district to use bio-fertilizer as to improve soil and environment in that area.

Discussion

As a research-funding agency, the Thailand Research Fund influences the agricultural extension projects by applying the community-based research concept and process. The top-down training process has changed to the participatory process by which all stakeholders have participated in the decision making process of identifying thematic concerned, research questions, objectives, expected output, action plan and so on. Through such learning process, the farmers have not been interviewed by the external extension workers, but actively participated as the core researchers who share ideas and experiences. Their self-confidence were step by step developed and increased. Similarly, extension workers who used to be the decision-makers have learned to listen to other core researchers, farmers in particular in order to collect data and information which has helped the decision- making more effectively.

What makes the agricultural extension project applying CBR concept and process, differs from the mainstream agricultural extension approach is the research question which all stakeholders hardly think of prier to the implementation of mainstream training project. The research question then becomes a useful tool for the TRF coordinator to encourage the core researcher to properly design action plan of each six-month phase.

Participation in action through series of action during the learning process has provided the core researchers, farmers and extension worker an opportunity to critically re-think about their belief and concept of agricultural extension and to improve their skills. For example, the core researchers have developed an ability to utilize information especially from critical reflection to synthesize the conditions contributed to the output of each activity, which then has been used to design how to adjust the next activity to be more effectively. Furthermore, they have improved such skills as data collection and analyzing, facilitating productive meeting, workshop and action of problem-solving activities. Such concrete usefulness has enabled the core researchers to change the way they defy evaluation to be more positive as an "on going learning process for improvement". Through such learning through practicing process, the core researchers have little by little changed their belief and ways of solving problems which is based on information and knowledge, rather based on feelings alone.

The success of each activity, supported by the well-planned action based on information, has greatly contributed to an increase of the confidence of the core researchers and those who have got involved in the CBR process. More often, some CBR activities did not succeed as expected, the core researchers would then synthesize the conditions which contributed to the failure, then adjust the next activity in a proper manner. Being positive thinkers, they consider the failure as the lessons learned, which can be used to warn the whole group to be aware of. The utilization of information has helped the core researchers to generate new knowledge which can be used to plan the next activity.

The importance of core researchers who become the decision-makers while undertaking research, rather than being researched by the external researchers as occurred in the past is now realized. They also realize that learning is coming from practice. It is this crucial condition that they have always confidently stated to the visitors who come to share experience and lesson learned. For local villagers, effective ways to disseminate experience can be facilitated through dialogue and practices. In this aspect, the CBR project has changed the agricultural extension project to be a continuous learning process by which core researchers, particularly farmers, are main actors, instead of passive participants.

An other crucial condition which contributed to the effective application of CBR concept and process in agricultural extension project is the facilitation process organized by the TRF coordinator. Another important condition is that financial support from the Thailand Research Fund to the CBR projects. According to CBR projects discussed above, it is important to note that agricultural extension workers have actively participated in the CBR projects because these projects dealt with their existing jobs or task which resulted in term of "they can manage existing jobs or task better". Moreover, agricultural extension workers have developed new training process to be "on the job training" process, developed new way of generating knowledge, skills and tools suitable for continuous learning process.

Lessons Learnt

Begin with the job that researchers can manage, emphasize on "task" As CBR concept and process is new to farmers and agricultural extension workers, they should be facilitated to begin with the job that can manage, emphasize on "task". Moreover, they integrate task with other partners in community they worked with.

Practice using of tools According to CBR concept and process, agricultural extension workers should be facilitated to practice using of tools so that they would then become new facilitator who can further facilitate farmers.

Working as peer to peer Considering examples of applying CBR project in agricultural extension project, discussed above, all stakeholders have been working as peer to peer based on their expertise and experiences. They become friends of various partners, yet using standard criteria participatory developed by the whole group.

Self-critical learning For those CBR projects discussed above, it is clear to note that best wider impact comes form starting small, exploiting its comparative competence in self-critical learning through developing, improving and spreading new approaches and methods learn from the NGOs' and GOs' experiences, and then passing that on to others. Quality assurance has been sought by stressing self-critical awareness and self-improvement which need "ready to learn" members.

Agricultural extension needs flexibility and adjustability: The dynamic, diversity and complexity of local community suggest that action plan of area-based research project has to be flexible and ready to be adjusted according to changes. Thus, extension worker has to be opened for such changes. TRF financial management system is also flexible enough to handle such unpredictability.

References

Opatpatanakit, Avorn, Lerttithikun, Saowaluck, Timkasikum, Wattanapong, Krairerks, Supapun and Wongjun, Dueanpen. (2004). Partnership in Educational Reform: Collaboration of Local Community, University and Research Funding Agency. Paper presented at the International Conference on EDU-Com 2004 New Challenges for Sustainability and Growth in Higher Education: The Second International Conference on Collaboration in the 21st Century. 24-26 December 2004. Sofitel Raja Orchid Hotel, Khonkhan, Thailand, 24-26 December.

Opatpatanakit, Avorn, Sumpun Chaitep and Nat Voryyos (2004). Empowering Mae Hae Noi Villagers to Manage Changes to be more self-reliance. Proceedings of The 11th Tri-University International Joint Seminar & Symposium 2004 on *Role of Asia in the World - Population, Food, Energy and Environment,* 26-31 October, 2004. Chiang Mai University, Thailand.

Opatpatanakit, Avorn. (2004). Community-based Research as Alternative Methodology for Developing Knowledge-Based Community. Pp. 46-51. Fedro Zazueta, Seishi Ninomiya and Royal Chitrado. *Proceedings of the AFIT / WCCA Joint Congress on IT in Agriculture. 9-12 August 2004.* Bangkok, Hydro and Agro Informatics Institute, National Science and Technology Development Agency.

Opatpatanakit, Avorn and Prason, Patcharin (2003). Empowering Dairy Cooperative to Improve Milk Quality, and Marketing Management, and Strengthen Cooperative Network. Paper Presented at The Tri-University International Joint Seminar & Symposium 2003 on *Role of Asia in the World -Population, Food, Energy and Environment,* 18-21 October, 2003. Mie University, Japan.

Opatpatanakit, Avorn (2001). Facilitating Learning Community through Empowering the Locals. Paper Presented at the International Workshop on *Participatory Technology Development and Local Knowledge for Sustainable Land Use in Southeast Asia,* 6-7 May 2001. Suan Bua Resort, Chiang Mai, Thailand.

Opatpatanakit, Avorn (1999). Participatory Action Research: Alternative Methodology to Research for Sustainable Agriculture. Pp. 173-178. In Sumalee Suthipradit, Chumpon Kunta, Suphachai Lorlowhakarn and Jaturaporn Rakngan. 1999. Sustainable Agriculture: Possibility and Direction, *Proceedings of the 2nd Asia-Pacific Conference on Sustainable Agriculture,* 18-20 October 1999. Phitsanulok, Thailand.

Opatpatanakit, Avorn (1998). Research Management for Increasing Milk Production: Northern Coordinating team, the Dairy Farmers' Socio-economic Research Project. Paper presented at the *Fifth Tri-University International Joint Seminar and Symposium 1998 on Role of Asia in the World Population, Food, Energy and Environment.* 26-29 October 1998. Chiang Mai University, Thailand.

13

Partnership between Research Institute and NGO to Alleviate Poverty

Nagaratna Biradar,
K.Sridhar and Arun Balamatti

Globalization of Indian economy during 1990's brought tremendous improvement in foreign investment in the country. This has created employment opportunities as well. However, much of positive changes witnessed are in the urban area which is true even for any of the initiatives taken to uplift the economic condition of the country. Of its nearly 1 billion inhabitants, an estimated 350-400 million still live below the poverty line, 75 per cent of them in the rural areas. This figure reflects that India still has the world's largest number of poor people in a single country. Prevalence of this large scale poverty remains the most shameful blot on the face of India. The country with vast geographical area and agro ecological diversity is making efforts to eradicate poverty. Since independence various poverty alleviating programs have been implemented through five year plans. It would be incorrect to say that all poverty reduction programs have failed. The growth of the middle class, which was virtually non existent when India became a free nation in 1947, indicates that economic prosperity has indeed been very impressive in India. But the problem lies in uneven distribution of wealth. As a result poverty in India has been reduced by 10 percent over the last few years. The basic causes of poverty are illiteracy, a population growth rate by far exceeding the economic growth rate for the better part of the past 50 years; protectionist policies pursued since 1947 to 1991 which prevented large amounts of foreign investment in the country (Ref.www.indiaonestop.com).

Rural poverty, hence, remains our unsolved problem. Most of this poverty resides in the farm sector, where a large part of our population, making use of a large portion of our land area ekes out a living. The basic reason behind this persisting poverty is seen to be our "land use practices." With a growing population, changing life styles, in combination with aggressive modern technologies, the production base in farming – soil, water and bio diversity – is getting rapidly depleted. In the process, the natural environment too is getting extensively damaged. The evidence is everywhere. The food we eat, the water we drink and the air we breathe have become matters of concern. It is high time therefore; that we devise, diffuse and adopt alternative, eco friendly farming practices.

Many technologies have gone through the process of technology validation and release at all India and state levels. But what is expected in ideal situation is the immediate spread and take up of these technologies and therefore better realization of crop yields and farm income to the families. The technology adoption level, however, in the country is not encouraging and several methods are being tried to hasten the process. Partnering with NGOs to diffuse technologies was successfully attempted by the Southern Regional Station (at Karnataka) of Indian Grassland and Fodder Research Institute headquartered at Jhansi, Uttar Pradesh. This chapter shares the experiences of partnership between research institutes and NGOs in diffusion of technologies to alleviate rural poverty.

Partnership between research institute and NGO

Indian Grassland and Fodder Research Institute (IGFRI)

Indian Grassland and Fodder Research Institute located at Jhansi, the Northern Part of India, is established in 1961 to coordinate and collate research work on fodder crops at National level and disseminate knowledge in this field. In the year 1987, it started southern regional station at Dharwad, Karnataka to address the farmers' problems through technology generation on forage crops, grasslands, silvi and horti pastoral management and forage seed production. This Regional Research Station though developed fodder technologies for small dry land holders for 10 years (1987-96) faced with a peculiar problem in reaching them. The land owned by these farmers are small and would be cultivated with grain crops as producing food for the family is always the main concern. Though livestock keeping is practiced, feeding with crop residues would suffice in a subsistence mode. The efforts made by the organisation by linking with various extension agencies to promote dry land fodder technologies bounced back with the message that none of the small farmers came forward to spare land for fodder cultivation. The niche areas, other than main land, required to fit in these fodder technologies were explored by field visits and discussions with various clients. The farm bunds laid in between and around the fields of small holder, embankment of farm ponds, the available land in between tree rows, the portion of land left fallow for reasons like gravelly, saline/alkaline etc. could be utilized to promote scientific cultivation of improved grasses and legumes to enhance the availability of quality fodder to these farmers. Livestock though reared for subsistence by small holders, is identified as one of the potential sources to improve the livelihood of these farmers. Because livestock is less prone to vagaries of weather, market for milk/meat is growing, the price fluctuations for these commodities unlike for grains is very marginal, besides farmers in the region traditionally have required skills to rear them. The rearing of livestock on crop residues alone however did not yield profit for them. Though milch animals during the period of lactation are fed with concentrate feeds this works out costlier reducing the profit margin. Hence encouraging this sect of farmers to cultivate improved fodder crops was felt essential to improve their livelihood options. It was also felt that this task of introducing entirely new fodder crops to the existing cropping system requires consistent effort, which could initially be achieved by tying up with some development agencies based in rural areas. The experiences of these efforts later could be scaled up by utilizing the various systems of formal/ informal extension network of the Deccan plateau. Therefore fodder technologies generated through a decade of research at Dharwad regional station were promoted effectively by partnering with BAIF Institute of Rural Development-Karnataka (BIRD-K).

BAIF Institute of Rural Development-Karnataka (BIRD-K)

BIRD-K is a sister organisation of the Bharatiya Agro Industries Foundation (BAIF) and is professionally managed by a team of multidisciplinary experts, under the guidance of the Board of trustees. Established in the year 1967 at Uralikanchan near Pune, it started with livestock development activities to alleviate poverty. Since 1996, BIRD-K is working in 22 cluster villages of Surashettykop in Dharwad district and is promoting tree based farming in smallholders' farms. These villages frequently face drought and farming in these villages could be grouped under Complex, Diverse and Risk prone situation. BAIF has identified around 3000 BPL families under its World Bank aided project aimed to diffuse technologies for sustainable agriculture. Under this project, IGFRI introduced its technologies by partnering with BIRD-K

Partnership between IGFRI-BIRD (K)

The lessons learnt from top down approach to diffuse technologies value participatory mode of knowledge diffusion. Initially discussions were held among all the actors comprising scientists of IGFRI RRS (D), managers and field staff of BIRD-K and the farmers regarding the approach to be followed to diffuse forage technologies. All the actors especially farmers expressed that they would like to have exposure to the various improved grasses and legumes available in the research farm before going for their cultivation. Therefore, the process initiated in 1997-98 with a series of exposure visits of small and marginal farmers in small batches to the research farm of IGFRI Dharwad center, where all the technologies were available in the field. It enabled the farmers to see and experience the performance of the technologies/varieties (feeling the softness of the fodder leaves, tasting stem for sweetness etc.) in the field grown in poor and gravelly soils. Besides, the planned interaction with the scientists facilitated to clarify their doubts and apprehensions about the technologies. *Brachiaria decumbense* and *Stylosanthes hamata* were selected by majority of the farmers for cultivating in identified niche areas. This grass and legumes were chosen by the farmers due to their drought tolerance and soil binding characters. Nearly 100 farmers came forward to take up selected grass and legume on bunds and fallow land in the field. The required seeds and root slips were distributed to these farmers followed by arranging demonstration on planting and sowing methods to be followed in the fields of 5 farmers. After lapse of 1 month, again in the village participating farmers meeting was arranged where in important agronomic practices to be followed and seed collection methods were explained. Majority of the farmers though took up sowing due to improper care like timely weeding; controlling grazing till they attain required height the crop was finally successful in the fields of less than 10 farmers. Among them only 3 farmers produced seeds. Seed production helped these farmers for two reasons – primarily found more remunerative vis-ḍ-vis other staple crops of the region and the second being generation of additional income from seed sale for the great demand within the state and elsewhere. The seeds produced by these farmers are in turn purchased by the BAIF for diffusing the same to other villages of the project area.

Growing of mixture of *Brachiaraia decumbens* and *Stylosanthes hamata* on bunds however got the attention of many neighbouring farmers. The newly formed bunds under the World Bank aided project by BAIF were very well stabalised wherever grass and legume mixture is grown as compared to other bunds left without vegetation. Also continuous availability of fodder attracted them. Groups of interested farmers were again exposed to fodder technologies in IGFRI farm and were supplied with seeds and root slips. Scientists also regularly visited their farms to keep up their motivation and when ever found essential all the actors met and discussed. Meanwhile new cultivars of Brachiaria and Stylosanthes found suitable to dry region on research station were also introduced, which were taken up by the participating farmers without any initial hesitation. They appreciated the new cultivars due to higher production of biomass and seeds.

The partnership initiated as early as 1997 resulted in spread of fodder technologies in more than 20 villages covering 300 and odd farmers. In January 2005, one day workshop of all the actors was held to analyse how the introduction of fodder technologies helped to strengthen the farm livelihood. The actors apart from IGFRI and BAIF participants comprised farmers who are successfully growing fodder crops as well as those who discontinued or partially adopted the technologies. Two successful cases were discussed in this workshop. Both the cases revealed that due to increased availability of quality fodder there was improvement in the health and conception rate of cows. This has lead to increased herd size.

Case of Mr.Mallappa Hakalad - meeting fodder and feed requirement

This particular farmer owns less than 1.5 ha of dry land having poor alfisols and 2 buffaloes, a cow and a pair of bullocks. In his field, five field bunds laid to contain soil erosion are planted with improved grasses. *Brachiaria decumbens* is planted on the top portion of the bund and lower reaches he planted *Pennisetum* trispecific hybrid. In between *Stylosanthes hamata* is sown. He has also sown *S. hamata* in small patch of his land for seed production. Cultivation of improved grasses and legume has helped him to stabilize bunds and harvest quality fodder. Everyday he cuts 70 kg of green fodder for his livestock. In Kharif, green native grass will be available to feed his cattle. During such period the harvest of stylo is dried, threshed and stored as hay to utilize during lean period. Similarly he prepares *Subabul (Lucena lucoecephala)* leaf meal from the trees planted around his field. Like this on an average he gets 8 bags of stylo hay and 16 bags of Subabul leaf meal, sufficient to utilize for 10 months for 12 cattle heads. As a result his expenditures on cattle feeds considerably decreased and animal health improved.

He also collects seeds of grasses and legumes, out of which he has earned Rs.2500 (US $ 55). Hay of *Stylosanthes hamata* and *Subabul* is mixed with Paddy husk while feeding. This saves him 10 q of paddy husk worth Rs.6000/-. Because of improvement in fodder availability and utilization he expanded his herd size and breeds. Presently he owns 4 upgraded buffaloes, a jersey cow, a pair of bullocks and 5 young buffaloes. Before interventions, he earned Rs.150/- per week by selling milk that has risen to Rs.750/- (US $ 17) per week, a five-fold increase in the earning. Apart from that most significant thing is increase in household consumption of milk and milk products by the family.

Case of Nagappa Adargunchi- earning from weak to healthy animals

This farmer owns half a hectare of dry land. Before interventions this land was given for share cropping and he with his family worked as wage laborers. The farmer now cultivates his land. He has grown Sapota and Mango trees and laid bunds inside the fields. On the bunds, *Brachiaria decumbens* is grown and *Stylosanthes hamata* in the inter-space of tree rows. Due to non availability of sufficient space to keep livestock, he resorted to the system of buying weak animals for bottom price from the *shandy (local market),* feed them properly making use of harvest from improved grasses and legumes. He also planted fodder trees Subabul and Sesbania around his farm, the loppings of which are utilized as fodder especially in summer season. Once the animal attains the desired weight and good health in about 4-6 months it is sold at higher rates usually 3-4 times the rate at which the animal is bought. If the animal conceives during the period of feeding, it is sold at the last stage of pregnancy to obtain maximum price. Commonly, he buys at Rs.4000 (US $ 90) and sells at the rate of Rs.14000-16000. To feed the animals, he depends entirely on fodder grown in his field. This farming system requires more skill in bargaining. Additionally due to the availability of cow dung he resorted to vermi-composting, the extra compost left after putting in his field is sold through BAIF. A small family is able to get engaged in his farm, apart from good earning. The addition of fodder in his farm enabled him to add livestock component in his farming system. The cow dung from livestock is utilized primarily for bio gas generation for cooking and the bio gas slurry is used for vermi-composting. The input-output chain of different components of farming system is not only apparent, but also sustainable as the use of external inputs is negligible. The village of this farmer faced three years of severe drought and this drought indeed benefited him. Distress selling of livestock during drought helped him to buy good breeds of animals at very low prices.

He fed these animals well in this period and allowed to calve. As a result he multiplied the animals during drought and sold some of them for higher price in the beginning of monsoon, the period when many farmers buy good animals.

Sample of 90 farmers comprising early and late adopters of fodder technologies were surveyed in 2005 to study the impact of fodder technologies on farm income and the result indicated that fodder crop contributed to the tune of 5.43 per cent in terms of sale of seeds and fodder to the total income and together with the livestock contribution was 21.29 per cent. Study conducted by the BAIF after 8 years of project intervention revealed that more than 80 percent of 3000 below poverty line (BPL) families are now above this line at the end of the project due to the transfer of sustainable technologies.

This partnership of research institutes and NGO which is different from the extension system prevailed during green revolution period is enabled to successfully introduce fodder technologies which otherwise was challenging and difficult.

Experiences of partnership of IGFRI and BAIF to introduce fodder technologies in the dry land small farmers

First benefit of partnering with NGO enabled the research station to straight away bank upon the rapport built by NGO with local farmers saving considerable amount of time and other resources. NGO had already done the bench mark survey of the area and hence IGFRI benefited to gain initial knowledge on the prevailing farming system of the region. The bench mark survey clearly indicated the families living BPL and IGFRI was looking to introduce technologies in small holder farming system. This complemented in selection of right target group.

This partnership is of significance as it dealt with those technologies which did not receive good acceptance from farmers through formal mode of extension practiced by the state department of agriculture. Also the partnership resorted to the participatory mode of technology transfer. The technologies were not thrusted on them instead farmers were allowed to select the fodder types which could fit well into their resource base. The research farm had number of grasses and legumes giving higher biomass than those chosen but would need more inputs. Farmers restricted to those which can withstand drought, yield reasonably good quantity and also got additional character of soil binding. Many of these farmers under the directions of BAIF had taken soil and water conservation measures mainly bunds and farm ponds in their farms. They selected *Stylosnathes hamata*- a legume crop along with grasses for obtaining quality fodder and also to produce seeds which are in high demand as discussed with the IGFRI. The beginning of this kind of extension enabled to set in motion those technologies which can fit well into the resource base of the end user.

Farmers while selecting fodder species resorted to various techniques to choose. Touch and feel the plant for softness, breaking the stem for texture/fiber, taste it for sweetness and observe the plant for pubescence were the four techniques which were followed by farmers of the entire batches visited IGFRI research farm. This indicates that farmers' decisions to select technologies are based on certain criteria and purpose (in this case fodder and soil conservation). These criteria might not have been perceived by the researcher unless and otherwise he/she interacts with the end-user periodically. Farmers' simple techniques to assess the quality of fodder crop as well are applauded, while scientists opt to complicated techniques to make certain simple decisions in the process of technology development. This indeed helped the scientists of IGFRI to learn farmers' decision methods and use some of these techniques while screening large number of plant species.

The partnership empowered the farmers by giving technical know-how of cultivation of these new crops which were totally new to the existing cropping system. The reinforcement of knowledge on cultivation practices by demonstrating in their fields by the scientists facilitated to take up timely operation for good establishment of fodder crops. During demonstration while scientists concentrated exclusively on content of the message, the NGO partner took care of fixing convenient field for demo, time and date by consulting farmers and ensuring that majority of them get benefit by attending it. This is of course the major advantage of partnering with NGOs in the process of technology development and diffusion. The flexibility in working time and style of NGO which is lacking in public extension system is certainly advantageous in such partnership. Time of farmers' availability in this region is not matching with the time followed by the public sector. NGOs, especially BAIF field staff reside in the target villages it self. This enabled them to approach the target group during the convenient and receptive hours of farmers. The research institute, as a result, benefited to ensure continuous flow of communication.

It would be impossible for the researcher to visit the field of farmers as often as the NGO partner. Field staff of BAIF placed in different villages visited their fields and observations made by them along with the farmer's were in turn shared to the researcher. Initial period of crop establishment is very crucial especially in case of *Stylosanthes hamata* and this was made clear to the field staff of BAIF. The planned visit of these field staff to the farmers' field was very critical for the success of introduction of fodder technologies. Because technology if failed in the initial stages, makes it obstinate to gain farmers' confidence further.

However, during the initial period of this partnering NGO also was little aware about improved fodder crops. The scientists thus made frequent visits to the NGO and farmers' fields during first two years. This helped for successful percolation of technologies in the farming community, coupled with making NGO field staff to understand minor intricacies of fodder technologies. For instance in case of *Stylosanthes hamata* only 40 per cent of the seeds sown would germinate in the first year and another 60 per cent remain dormant and germinate in the second year. This peculiar feature of the crop is usually misunderstood to the quality of the seeds as is judged in terms of germination percent. The poor stand of the crop in the first year was explained clearly by the researcher to the NGO field staff and in turn to the farming community at large. As a result technology was taken in true spirit both by the NGO and the farmer. Analysis of strengths and weaknesses of NGOs has clearly brought out that while NGOs are good in social dimension of development effort but are poor in technical know-how. Other way round, Government extension agencies are strong in technical side of the development. NGOs by partnering with research institutes could get an opportunity to fill this gap and this is one of the distinguished benefits they would get out of the partnership.

During first two year the NGO field staff learnt about several fodder crops and cultivation practices. Subsequently, they were able to solve many of the field problems. When the research institute contacted NGO managers in the first year, the latter showed hesitation mentioning that they are not aware about these crops and their importance. Eventually due to the partnership they learnt about these crops and realized their importance in sustainable farm practices. Presently this NGO in the content of its training programs gives enough space and importance to fodder technologies. Field staffs associated with the research institute explain the participants about the fodder technologies with very little help from the scientists. The gain and spread in knowledge on fodder crops thus was possible by partnership.

Besides the target oriented approach of NGO staff facilitates to closely observe the performance of the technology in the farmers' field that is essential in the initial years to gain the confidence of the farmers on new technology. This perhaps is difficult to expect from the public extension system.

Government extension workers are being stigmatized as the carriers of subsidies, free seeds and inputs to the farmers. Farmers are hence more interested to avail these facilities than genuinely getting benefited from them the technical know-how. However, NGOs are free from such stigma and perceived as committed workers to serve the rural poor. As a result they are received very well by the rural community and research institutes which generally discourage subsidies can bank upon this impression of the NGO to transfer the technologies.

Essentially research institutes are expected to develop technologies which are problem oriented and fit well into the resource base of the target group. Partnering with NGO helps the scientists to get sensitized about the field problems and concentrate on developing need based technologies. This is in turn the splendid benefit the research organisation can get benefit by this partnership.

This symbiotic relationship hence worked out very successful in the partnership between IGFRI-RRS (D) and BAIF, Dharwad as was evident from the case studies mentioned in the earlier part of the chapter.

Regarding scaling up of such efforts the fields of these two case study farmers now attract a wide range of development agencies starting from extension unit of State Agricultural Universities, State Departments of Agriculture/Forestry/Animal Husbandry, reputed NGOs like AME, Bangalore, Agriculture Science Foundation, Hulkoti, MYRADA, Bangalore etc. and from the neighboring states as well, gesturing the process of scaling up.

Conclusions

The research set up alone cannot create an outreach adequate to meet the demands of the situation. NGOs should not, however, be seen as the extended arm of Government. In the era of changing development scenario where participatory development approaches are being accepted by everyone, NGOs and research institutions, along with various other development actors can give the required fillip to development process through close collaborations. Collaborating with the relevant organisations of the repute should be based on the problem in hand and the goal set for. Previous experiences and resource base of NGO should be considered while collaboration. Moving from Green revolution to Evergreen Revolution indeed requires concerted action. In this direction partnership between research institutes and NGOs is crucial in the effort to eradicate poverty through technology introductions.

PART 4

INCREASING MANAGERIAL ABILITIES OF FARMERS AND THE USE OF ICT

14

The Role of the Farmer Field School in the Transition to Sustainable Agriculture

Elske van de Fliert

Introduction

The introduction in Southeast Asia of the Green Revolution technologies, such as high-yielding varieties, chemical fertilisers and pesticides, was typically managed through top-down interventions by often newly established extension services. It were exactly the negative effects of these technologies occurring over time, like the deterioration of soil fertility, decline of agro-biodiversity, development of resistance and resurgence of pests leading to pest outbreaks, and increased social inequity, demonstrating that for the introduction of sustainable agricultural practices and approaches a different extension paradigm is needed.

A shift to sustainable, ecologically sound, economically viable and socially just agriculture (Reijntjes *et al.*, 1992) requires knowledgeable and skilled farmers who can make locally specific decisions, adapt technologies to their particular needs and conditions and resolve conflicting interests, rather than the promotion of a standard set of recommendations and technology packages to be implemented unquestioningly. Hence, farmer education and social learning (Rōling and Jiggins, 1998), rather than technology transfer, is what is needed. The farmer field school (FFS) evolved as a model to address the needs for this paradigm shift from a series of experiments in rice pest management in the Philippines (Adalla and Rola, 1986; Kenmore *et al.*, 1987), and was first piloted for rice integrated pest management (IPM) on a significant scale in Indonesia in 1989 (Van de Fliert, 1994; Matteson *et al.*, 1994). Soon, IPM FFS programmes were carried out, initially with FAO support, in rice cultivation in twelve Asian countries, and gradually branched out to vegetables, cotton and other crops with increasing support from other organisations.

From the mid-nineties onwards, the experiences generated in South and Southeast Asia were used to help initiate IPM FFS programmes in other parts of the world, including Africa, Latin America, Eastern Europe, Central Asia and the Pacific. Curricula for new commodities and issues, such as livestock and HIV-AIDS, have been developed (see ILEIA, 2003, and CIP-UPWARD, 2003, for a varied array of reports of FFS programmes), while local adaptation and institutionalisation is encouraged. A variety of follow-up activities to FFS have been designed and implemented in various countries as to address the need for sustained learning, problem solving and collective action among FFS alumni groups. Community IPM, participatory action research and IPM clubs are amongst these (Pontius *et al.*, 2002). At present, farmer field school programmes, in various modes and addressing a wide range of development issues, are being conducted in 78 countries worldwide (Braun *et al.*, 2005/in prep.).

The effectiveness of the FFS as a farmer education model and social learning platform has received ample recognition (e.g., Rōling and Van de Fliert, 1994; Godtland *et al.*, 2003; Van den Berg, 2004). Most credit is given to its capacity to increase knowledge and enhance skills, and to serve as a platform for collective action. FFS programmes have, however, been criticised for being costly and not always achieving substantial economic impact (Feder *et al.*, 2003). The question is, however, what is desired on the long term – an immediate economic improvement or long lasting capacity of farmers and communities to manage the agroecosystem in a sustainable way, considering both economic and environmental, human and social indicators.

This chapter addresses the potential, requirements and limitations of the farmer field school as an approach to promote sustainable agriculture. It first outlines the principles and practices of the FFS approach, and explains why it is a suitable model for farmer education. In the next section, the various applications of the FFS are discussed and the implications for designing and managing FFS programmes. Through a description of cases (Vietnam and Indonesia), a review is given of experiences and issues relating to the institutionalisation of FFS as a mainstream extension approach. Last, some lessons learned will be summarised and conclusions for the prospect of the FFS approach drawn.

The farmer field school as a platform for learning, problem solving and collective action

Sustainable development can be defined as a change process to improve people's livelihood on the short and long term, taking into consideration socio-cultural, economic, environmental and political aspect. In the context of agriculture this implies a need for farmers to become knowledgeable and skilled managers of the agroecosystem, who obtain optimal output of their enterprise at minimum input while safeguarding the environment as to support present and future production, and maximising their share in the market. Within the context of most Asian countries, the majority of farm landholdings are small, financial capacity of farm families is often limited and labour is relatively cheap. Agricultural technologies are generally developed under research station and model farm conditions and do not exactly suit the highly variable conditions that the majority of smallholder farmers are facing. Hence, farmers need to be able to adapt information and innovations conveyed to them by a variety of sources, such as extension and development agents, mass media and input retailers, to the agroecological and socio-economic conditions prevailing on their farms. Additionally, empowerment has become increasingly crucial for farmers who live and work in a world where they face a variety of contending forces related to technology, politics, world markets, and society, which can marginalize farmers if they are not proactive. Farmers need to be able to make their voice heard as the need for sustainability in agricultural development become more urgent.

For the sustainable management of the land which emphasises the establishment and conservation of an ecological balance in the field, smallholder farmers are interdependent, since ecological processes are not restricted to the artificial boundaries of farms with crop fields often smaller than half a hectare. As a result, the cultivation practices of one farmer can either benefit or constrain the ecological conditions of their neighbours' fields. Farmers do, therefore, benefit from collectively making decisions and taking actions relating to issues that are of importance in this respect. Some examples are the conservation of natural enemies by eliminating the use of broad-spectrum pesticides, large scale and continued suppression of rodent populations, and the prevention of ground and surface water contamination by pesticides, inorganic nutrients and disease inoculum. These conditions can not be successfully achieved when endeavoured as an individual.

An extension model that accommodates the requirements of enhancing knowledge, critical skills and collectivity should be characterised by farmer and community- centred, rather than technology-centred, approaches. Farmers need to be capable of sifting available information and deciding what is valuable for them and worth trying out. Underlying basic knowledge about, for instance, ecology and crop physiology and strong skills relating to observation, experimentation and farm economic analysis will help them to make decisions and develop a farm management system that suits their conditions. Obviously, new information and innovations need to be made available to them, too, but not as a recommendation to follow blindly, but as one of the many options to choose and adapt from. Farmer education is therefore a more suitable term to capture the type of interventions required to support farmers in the transition to agricultural sustainability, rather than extension – a concept that comes from the typical technology transfer tradition where innovations were developed by researchers and "extended" to farmers through extension officers, without farmers having a say about what they need or what is feasible on their farms. Consequently, extension officers become facilitators of a process entailing learning, problem solving and collective action, rather than agents promoting standard technologies.

The farmer field school (FFS) model claims to be an approach that responds to the abovementioned requirements. The FFS was designed based on the principles of non-formal education, which builds on the concept that adults learn optimally from real-life experiences through observation, experimentation and analysis under for them relevant circumstances. Additionally, it emphasises group learning as the basis for local problem solving – building on farmers' own experiences and experimentation – and further collective decision making and action.

An FFS engages a group of typically 25-30 farmers in season-long learning activities, which take place in the field or centre around concrete field problems. The FFS participants cultivate a chosen crop and conduct various experiments together on a learning plot. During weekly sessions of on average 3-4 hours, participatory learning exercises are offered that relate to the development stage of the crop and problems that emerge during that particular stage. Routinely, FFS participants observe the crop on the learning plot, draw up an agroecosystem analysis, and make decisions together what action is needed with regard to crop management practices. Additional experiments and special topics give further insights into plant physiology, soil and crop ecology, pest and disease management and socio-economic aspects of the crop enterprise. To enliven the FFS process and reinforce group coherence as to encourage communication, collaboration and leadership, group dynamics exercises are interspersed with the learning exercises. A heavy emphasis is placed on farmers sharing their ideas and knowledge, and presenting and discussing findings of learning activities.

This appears to be highly empowering for farmers who are used to be seen as the elements at the end of chain that are supposed to be following instructions from above in the top-down technology transfer tradition.

Farmer field schools can only be implemented successfully if facilitated in a way that is consistent with its design. Trainers should facilitate a discovery learning process instead of giving lectures and instructions. The training of trainers is crucial to FFS programmes for effective implementation. In most instances, FFS programmes are conducted through government agencies and either agricultural extension officers or technicians are assigned and trained to be FFS facilitators. These officers have mostly learned themselves through directive, top-down education systems and have likely been involved in top-down farmer extension efforts before. Thorough training of trainers through participatory and experimental learning methods, just as the FFS employs itself, is needed before anyone can be expected to facilitate participatory and experimental learning activities. Typically, training of trainers events in FFS programmes are season-long as well, as to allow the to-be facilitators to go through all crop cultivation practices and learning exercises themselves that they will take the farmers through later.

In addition to cadres of qualified facilitators, and like any other form of intervention aiming at sustainable development, programmes applying the FFS approach require support mechanisms for effective and efficient implementation. FFS programmes are resource intensive considering time and money investment, albeit with a promise of sustained impact, and hence require capability and commitment from the implementing agencies to allocate these resources. Considering the importance of continuity in learning, problem solving and collective action after a season of FFS in order to achieve sustained impact, local government support should also be established. Consistency with simultaneous interventions, or at least the articulation of possible conflicting interventions, is another area that requires attention when FFS approaches are being institutionalised.

Several challenges and issues emerge with the scaling-up and diversification of FFS programmes. Quality maintenance of both content and the facilitation process in both training of trainers events and farmer field schools is a main concern for FFS programme managers. Appropriate adaptation of content and methodology to changing conditions or diversified uses of the model is another. A solid internal monitoring system needs to be in place to guide managers and facilitators. FFS facilitators regularly need to be updated regarding innovations relevant to the content of the FFS , for instance new biological control methods that were developed in the case of IPM FFS, and develop new experimental learning exercises for new pieces of information or innovations. The sections below illustrate how some FFS programmes have dealt with these issues.

Applications of the farmer field school

Originally, the farmer field school was designed for rice integrated pest management (IPM). The FAO supported rice IPM programme operated in twelve countries in South and Southeast Asia until 2002, and has been expanded by national programmes, other donors (especially the Danish government) and non-government organisations in all of these countries and beyond. The main problem addressed was simple –

brown planthopper outbreaks as a result of injudicious pesticide use - and so was the remedy – avoid using broad-spectrum insecticides to conserve natural enemies. In FFS, farmers learned the basics of insect life cycles and predator-prey relations, how to observe pest insects and natural enemies, analyse the balance between the two categories and make a decision whether or not a control practice is needed. To demonstrate that pesticide use would rather aggravate the insect problems than cure them, the FFS contained a standard experiment in the learning field with two treatments: calendar-based pesticide application (the farmer practice or FP plot) and pest management practices based on the agroecosystem analysis, mostly implying no action at all (IPM practice plot). The FP-IPM experiment became part and parcel of the FFS design. Impressive reductions in pesticide use were made in rice production in the various countries (The FIELD Alliance, 2002).

Over time, more and more topics have been addressed in the rice IPM FFS, ranging from seed selection to disease and nutrient management and in many places even marketing issues. Although many of these topics would lend themselves perfectly well for simple, problem-specific trials, generally facilitators would stick to the FP-IPM concept. Due to the complex of changes introduced in the IPM treatment as to address the range of improvements that farmers can make to grow rice more sustainably, the trial is no longer a simple two-treatment experiment, but rather a demonstration of two different "packages". There is a danger here that farmers do not get to know what exactly the differences are in the individual practices these two "packages" encompass, and as a result cannot draw clear cut conclusions as to why IPM might be better than farmer practice. Ideally, new simple experiments should be designed and conducted in the FFS learning field, that focus on one issue of concern at a time, for instance the nitrogen-potassium ratio in fertiliser application. This would, however, require upgrading of all FFS facilitators through refresher courses, which is a timely and costly investment that most (particularly large scale national) programmes cannot accommodate and afford.

Moving into FFS for other crops, which extensively happened after the initial years of success of rice IPM FFS, has made the picture even more complicated. For instance, in vegetable IPM not only is the post and diseases scenario more complex, but also are we dealing with a variety of crops from several different families that are often grown in mixed or intercropping systems. For vegetable IPM to be effectively implemented, the importance of farmers getting involved in location-specific problem solving and collective decision making and action is even more important than for any other cropping system. There are ample arguments that one season learning about vegetable IPM in an FFS is not enough for farmers to get beyond the basics of informed decision making and implementing some alternative practices effectively. To become confident vegetable IPM implementers, farmer groups in several Asian countries have been engaged in multiple season activities, some of which are reported in the section below. This, however, increases the intervention cost per farmer, which according to some criticisms is already high with single season FFS, and large scale implementation may become unfeasible. Nevertheless, this is where some real changes can be made with regard to a transition to more sustainable agriculture, since the level of chemical inputs is very high in commercial vegetable production in most Asian countries, leading to high chemical residues and not uncommonly rejection of products at international markets. Particularly in the vegetable sector, where farmers seem to receive most of their information from chemical input retailers, a counteracting force of empowered farmers who know what they want is needed to balance the commercial drive of input producing companies to promote their products. The farmer field school provides a platform that can accommodate these processes, but adjustments from the original model will need to be made. Additionally, commitment from implementing and funding agencies is needed to do a quality job as to guarantee quality impact.

Other areas that FFS became popular for as a model for introducing change in rural communities during the past couple of years are livestock (Minjauw *et al.*, 2004; Dalsgaard *et al.*, 2005) and HIV/AIDS (Sokunthea, 2002), under the names of Farmer Livestock Schools and Farmer Life Schools, respectively. These approaches build on the underlying principles of non-formal education of the FFS, but employ a range of adapted methods and activities to suit the specific needs of these content areas.

A major lesson learned is that when diversifying the application of the FFS model, both with regard to content and geographical areas, rigorous rethinking is needed about the actual implementation design to achieve consistency between context, content and methodology. No single model can be used as blueprint for application in other content or geographical areas, but always needs to be adapted to suit the specific conditions, needs and opportunities. Interestingly, this need for adaptation has been addressed more adequately in content areas that are remote from the initial rice IPM FFS rather than similar cropping systems, as well as in geographical areas remote from Indonesia where the rice IPM FFS was initially developed. Over time, it has become clear that making adjustments on the way is more difficult after heavy investment has been made to train trainers through intensive season-long preparation. FFS programmes should have a solid internal monitoring and evaluation system in place, together with feedback mechanisms, and deploy the flexibility to adjust methods and activities along the way.

Institutionalisation of FFS

As can be extracted from the descriptions above, the farmer field school model is in principle complementary to conventional agricultural extension services employed by local governments, NGOs or community-based initiatives. It works on educating farmers to become better decision makers, who can then better choose from the basket of options provided to them through the various sources of information. It can, however, also be conflicting. The establishment of groups of empowered farmers may not always seem desirable to local governments or leaders in the context of socio-political conditions in certain countries or regions. Additionally, some government extension programmes have quota for adoption of technologies, for instance new varieties, fertiliser types or herbicides. Extension officers' performance is evaluated based on, among other aspects, whether or not they achieved their quota, which can lead to biased advice to farmers. In other cases, as is the case in many developing countries, extension officers need to supplement their salaries to support their families, and doing this from agricultural input sales is the most obvious for many of them. These situations and their implications need to be considered when designing FFS programmes and determining where programme implementation units should be based, and who should be trained to become facilitators. This is vital in anticipating institutionalisation and sustainability of the approach. Two examples of how this was dealt with within national programmes managed by government agencies are elaborated below: Indonesia and Vietnam (Pontius et al., 2002).

In Indonesia, the country where IPM FFS was piloted and introduced first, not extension officers but pest observers from the provincial Crop Protection Centres were selected to become FFS facilitators. Main reason was that the role of extension officers consisted largely of input distribution, commonly implying gain of personal profit. Pest observers, on the contrary, were mainly involved in doing surveillance and reporting pest outbreaks. The IPM FFS programme was set up parallel to the conventional extension system, which indeed did result in farmers receiving conflicting message at times. In the course of the years, when the requirements for trainers exceeded the available pool of pest observers, an increasing number of extension officers became involved, however with mixed successes. Soon, the programme began to involve and train farmers as trainers under the umbrella of the Community IPM Programme, which was the last phase of the FAO supported rice IPM programme in South and Southeast Asia (The FIELD Alliance, 2002). This resulted in the establishment of the Indonesian IPM Farmers Association in 1999, facilitated by over four hundred farmer trainers spread across Indonesia who provided the leadership in getting this organisation IPM FFS alumni for up and running. This association became an independent body with its own funding and organisational structure, which to date is still active throughout all administrative levels in Indonesia society. Activities organised by the Indonesian IPM Farmers Association, conducted mostly with locally generated funds from district governments, range from traditional farmer training through FFS and facilitation of IPM research group, to the organisation of provincial or district farmer congresses where FFS alumni exchange experiences and plan activities In the period from 1989 until early 2005, more than 1.1 million farmers were trained in a variety of IPM FFS (Braun et al, 2005 in prep.). While the FAO supported Rice IPM programme terminated in 2002, and continuation of IPM FFS activities by the national government has become very limited, the Indonesia IPM Farmer Association continues to exist and support fellow farmers in the transition to more sustainable and profitable farming.

In Vietnam, at the time the FFS was introduced in 1992, no agricultural extension system existed yet. Hence, technicians from the provincial Plant Protection Sub-Departments (PPSD) and district Plant Protection Stations (PPS) were selected to become FFS facilitators. Efforts to engage the extension system that was established later have been limited (except for the case of livestock FFS), partly due to the segmented organisation and implementation of duties of the provincial Departments of Agriculture where both PPSDs and Extension Service are located. To expand the cadre of FFS facilitators, farmers became rather involved and were trained as trainers as part of the Community IPM Programme. Both local Farmer's Union and Women's Unions were represented in the training of farmer trainers events, providing a good base for FFS extension through these mass organisations. A major role for farmer trainers within the context of he Community IPM Programme was to organise IPM Clubs for FFS alumni through which FFS graduates would continue to do field studies and solve cultivation problems together. As opposed to Indonesia where through the newly established association farmer trainers began running their own programme independently, in Vietnam FFS activities are typically coordinated by government agencies. Farmer trainers are hired as co-facilitators in these government agency managed programmes. A variety of international NGOs adopted the FFS approach, as well, but the preferred mode was by hiring PPSD/PPS FFS facilitators rather than building up own trainer capacity in their constituencies. In the period from 1992 until early 2005, an estimated 930,000 farmers were trained in a variety of IPM FFS (Braun et al, 2005 in prep.). Although the majority of FFS in Vietnam have been funded through externally funded projects, mainly through FAO and its associated donors (the governments of Australia, The Netherlands and Norway) and through bilateral aid by the Danish government, a great deal of rice IPM FFS during the late 1990s were financed through local funding from provincial governments. Currently, vegetable IPM FFS are being mainstreamed into locally funded Safe Vegetable Programmes, whereas the FAO supported Regional Vegetable IPM programme (Phase II, 2003-2007) concentrates on training enough trainers to support these local programmes.

Scaling up and out through institutionalisation brings about the risk of diluting quality. However, experience has shown that adaptation of methodologies and organisational aspects to local conditions and needs and incorporation into locally supported programmes better guarantee a post-programme life of the FFS. The risk for no continuation when funding stops is larger when a programme is implemented in isolation. This is particularly true for programmes applying approaches such as the FFS that require substantial time and money investments.

Prospects for the FFS

Although the farmer field school model has already celebrated its fifteen birthday, the approach is still very relevant as a methodology or platform to contribute to the continued transition to more sustainable agricultural systems. Awareness about the hazardous effects of chemical agricultural inputs has generally increased among farmers, consumers and policy makers alike, which puts some pressure on farmers to minimise chemical use in agricultural production. Nevertheless, new issues have arisen that farmers have to keep up with. With markets expanding globally, particularly for those countries joining the WTO, farmers have to become increasingly knowledgeable about quality issues, marketing channels and certification systems. For smallholders, collective action in marketing might become a crucial mechanism to survive. Hence, continued learning, problem solving and collectivity, as is supported by the FFS, albeit with a changed content focus, remains important. As has always been the case, continuous adaptation of content and methods is needed to adjust to changing requirements and conditions, either those of the individual farmer or of the larger community. National programmes should try to avoid becoming dependent on external funding, but gradually mainstream the FFS approach in an appropriate format within their own government programmes, allowing for adherence to the basic principles of the approach and maintaining high quality, as to guarantee sustainability. With an increasingly larger group of increasingly smarter farmers, a meaningful transition can be made to agricultural systems that hold a promise for the next generations.

References

Adalla, C.B. and A. C. Rola (1986). Participatory Verification and Technology Generation of Location Specific IPM Technology in Calamba, Laguna: A Pilot Project. In: *Filipino Women in Rice Farming Systems.* U.P. at Los Baños:, IRRI and the Philippines, Institute for Development Studies: p. 41-56.

Braun, Arnoud R., J. Jiggins, N. Roling, H. van den Berg and P. Snijders (2005, Final Draft). A Global Survey and Review of Farmer Field Schools Experiences. Report prepared for ILRI. Wageningen, The Netherlands : Endelea. 88 p.

CIP-UPWARD (2003). *Farmer Field Schools: Emerging Issues and Challenges.* Los Baños: CIP-UPWARD, 434 pp.

Dalsgaard, J.P.T., T.T. Minh, V.N. Giang and J.C. Riise (2005). Introducing a farmers' livestock school training approach into the national extension system in Vietnam. *AGREN Network Paper* No. 144. 15 pp. http://www.odi.org.uk/ag-ren/papers/agrenpaper_144.pdf.

Feder, G.R. Murgai and J. Quizon (2003). Sending Farmers Back to School - The Impact of Farmer Field Schools in Indonesia. *Policy Research Working Paper* 3022. Washington D.C., World Bank. 48 pp.

Godtland, E., Elisabeth S., A. de Janvry, R. Murgai and O. Ortiz (2004). The Impact of Farmer-Field-Schools on Knowledge and Productivity: A Study of Potato Farmers in the Peruvian Andes. *CUDARE Working Papers* (963). Berkeley: University of California. 24 pp. http://repositories.cdlib.org/cgi/view-content.cgi?article=1061&context=are_ucb.

ILEIA (2003). *LEISA Magazine – Learning with Farmer Field Schools,* 19(1). 36 pp.

Kenmore, P., Litsinger, J. A., Bandong, J. P., Santiago, A. C. and Salac, M. M. (1987) Philippine rice farmers and insecticides: Thirty years of growing dependency and new options for change. In: Tait, J. and B. Napompeth (eds), *Management of Pests and Pesticides. Farmers' Perceptions and Practices.* Colorado: Westview Press: p. 98-115.

Matteson P.C., K.D. Gallagher and P.E. Kenmore (1994). Extension of integrated pest management for plant hoppers in Asian irrigated rice: Empowering the user. In: Denno, Robert F. and T. John Perfect, *Plant Hoppers: Their Ecology and Management.* New York: Chapman and Hall, pp 656-685.

Minjauw, B., H.G. Muriuki and D. Romney (2004). Development of Farm Field School methodology for small-holder dairy farmers in Kenya. In: Owen, E., T. Smith, M.A. Steele, S. Anderson, A.J. Duncan, M. Herrero, J.D. Leaver, C.K. Reynolds, J.I. Richards and J.C. Ku-Vera (Eds), *Responding to the Livestock Revolution - The role of globalisation and implications for poverty alleviation.* Nottingham, UK: British Society of Animal Science Publication No. 33, Nottingham University Press. p. 299-313. http://bsas.org.uk/downloads/mexico/046.pdf.

Pontius, J.C., R. Dilts & A. Bartlett (2002). *From farmer field school to community IPM: Ten years of IPM training in Asia.* RAP/2002/15. Bangkok: FAO Regional Office for Asia and the Pacific. 106 pp. http://www.communityipm.org/-docs/10%20Years%20of%20IPM/10Years-Main.htm.

Reijntjes, C., B. Haverkort and A. Waters-Bayer (1992). *Farming for the Future. An Introduction to Low-External Input and Sustainable Agriculture.* London: Macmillan Education Ltd. 272 pp.

Rōling, N. G. and J. Jiggins (1998). The ecological knowledge system. In: Rōling, N. G. and M. A. E. Wagemakers (eds), *Facilitating Sustainable Agriculture: Participatory Learning and Adaptive Management in Times of Environmental Uncertainty.* Cambridge, Cambridge University Press: p. 283-311.

Rōling, N. G. and E. van de Fliert (1994). Transforming extension for sustainable agriculture: the case of integrated pest management in rice in Indonesia. *Agriculture and Human Values* 11: p. 96-108.

Sokunthea, N. (2002). *Mobilization and Empowerment of Rural Communities Along the Asian Highway (Route 5) in Cambodia to Reduce HIV Vulnerability*. Report Project UNDP-FAO/RAS/97/202. http://www.communityipm.org/-docs/Farmer_Life_Schools.doc.

The FIELD Alliance (2002). *Community Integrated Pest Management.* http://www.communityipm.org/.

Van den Berg, H. (2004). *IPM Farmer Field Schools: A synthesis of 25 impact evaluations.* Rome: Global IPM Facility, FAO. 53 pp. http://www.fao.org/ag/AGP/AGPP/IPM/gipmf/en/03_topics/03a.htm.

15

A Framework For A Pluralistic Extension Agenda Under A One System One Program Approach

Samuel T. Mancebo

Introduction

The enactment of the Agriculture and Fishery Modernization Act (AFMA) in the Philippines is premised on the need to modernize agriculture. Modernizing agriculture must involve the pursuit of objectives that our Society currently values: food security, competitiveness, poverty alleviation and protection of the environment. Of the four objectives, however, poverty alleviation is the primordial and ultimate justification for the efforts to modernize Philippine agriculture. Agriculture and poverty are linked; most of the poor are in the rural areas. Thus, making agriculture more productive will help the battle against poverty in general (Agricom Report, 1998 p. 36).

The Reasons for modernizing Philippine agriculture

Paraphrasing further the Agricom's definition of

modern Agriculture, it says that three features are always present that characterize a modern agriculture.

They are as follows:

a) It is a dynamic society that welcomes innovation and change;

b) It is highly productive and competitive because it uses modern production and management technology; and

c) Its emphasis is on the enterprising tiller who exercises his right to choose what technology to apply, what crop to raise and when, and to whom to sell his produce to get the highest returns.

PHILOSOPHY

From the above features, the Philosophy of modernizing Philippine agriculture involves a change in attitude by the farmer toward modern technology to increase his productivity and income. At the very core of this definition is that the farmer is given the opportunity to choose from a range of options available to him or her. The choice must be an informed one, it must be a choice not forced by circumstances but by his or her own free will.

The emphasis on choice, on the farmer being transformed into an entrepreneur, necessarily implies that he has the required skills and assets, and has access to modern technology (whether in the form of modern equipment, roads, telecommunications and cellphones to access the market and technologies.)

Dominance of Small Farms

By the turn of the century, given a high rate of population growth (2.3%) combined with rapid urbanization and the agrarian reform making headway of breaking large estates with small farms, THE FUTURE PHILIPPINE AGRICULTURE WILL BE DOMINATED BY SMALL OWNER CULTIVATED FARMS. Productivity and innovative marketing approaches will be the preoccupation of government in Agriculture (Figure 1).

The predominance of small farms will predetermine the nature of technologies invented, products produced and institutions formed. Our system of education, our production and processing units and the delivery mechanism, will have to be modified accordingly and be governed after the imperatives of the small farms. Therefore, it will not be the land size that will matter but whether we can equip our farmers with the skills, assets and technology of intensive cultivation with awareness to maintain ecological balance in order to increase their yield.

Educating Farmers and Entrepreneurs

The ability of modern Philippine agriculture to compete in the global economy will be determined by whether or not farmers/fisher folks are equipped with the necessary knowledge, skills, and techniques. A modern agriculture cannot be attained without educating farmers properly. Their ability to appreciate and adopt a new technology hinges on their level of comprehension. Education plays an important role in making the farm efficient. He can grasp and process information more quickly and make wise choices from among a range of options. Hence, it is worth investing in education, training and extension because people are the sources of competitiveness in agriculture.

One of the components of the AFMA law was the establishment of a National Agriculture and Fisheries Education System (NAFES). The NAFES is an integrated system of agriculture and fisheries education covering both public and private institutions in all the three levels of Philippine education in the elementary and secondary schools, the post-secondary institutions offering non-degree technical vocational programs in agriculture and fishery and at the institutions of higher learning offering degree and post-graduate programs in agriculture and fishery.

One objective worth mentioning at the basic education level of the NAFES is to develop an outreach program where students, parents and the schools become instruments in affecting positive changes in the pupil's home and community. This extension objective is now being demonstrated in the Family Farm School Experience show cased as the community-supported Agriculture or corporate-supported Agriculture (3).

Box 1.0 Community-Supported Agriculture (CSA). It is a new idea in the Philippines,

Community-Supported Agriculture (CSA) is a new farming idea in the Philippines, although the concept started as early as the 1960s in Switzerland and Japan where consumers interested in safe food and farmers seeking stable markets for their crops joined together in economic partnerships.

Suzzane DeMuth of the Alternative Farming System defined CSA to consist of a community of individuals who pledge support to a farm operation so that the farmland becomes, either legally or spiritually, the community's farm, with the growers and consumers providing mutual support and sharing the risk and benefits of food production. Typically, members or "share-holders" of the farm garden pledge in advance to cover the anticipated farm operation cost, including the farmer's salary. In return, they receive shares in the farm's bounty throughout the growing season, as well as satisfaction gained from reconnecting to the land and participating directly in food production. By direct sales to community members who have provided the farmer with working capital in advance, growers receive better prices for their crops, gain some financial security, and relieved of much of the burden of marketing.

Although CSAs take many forms. All have at their center a shared commitment to building a local and equitable agricultural system, one that allows growers to focus on land stewardship and still maintain small yet productive farms.

Another form of an emerging innovation of turning the graduates of the NAFES into entrepreneurs and viable partners in modernization in the Young Farmers Program. The program's rationale is as follows:

"The future of the country's agriculture rests on the hands of farmers who are willing to practice it. Currently the average age of the farmers is 67 years. They are tired and feeble and are no longer receptive to change nor to modern technology. Their children who are supposed to replace them prefer to work in menial jobs in the cities or end up as domestic helpers or care givers abroad.

The same is true with those who labored to finish agriculture and fisheries in college. They are afraid to venture into agribusiness owing to their lack in self-confidence and the high rate of failure among small and medium enterprises (SMEs) in the sector. Despite recent developments showing agribusiness as profitable, these failed to provide the motivation to the youthful graduated to pursue careers in agriculture and fisheries.

The Young Farmers Program (YFP) aims to instill among young graduates the dedication and commitment to the development of agriculture and fisheries. It endeavors to equip them with appropriate technical skills, provide production and marketing support and assist them in capital formation with the hope that they may contribute to the nation's food security, generate employment and help the sector face the challenges of globalization.

Box 2.0 YFP to help finance more agribusiness projects

The Young Farmers Program (YFP) is now accepting the processing of viable agribusiness project proposal of applicants who will be scheduled to attend the intensive training on business management for the 4th batch of participants to the program.

The YFP, which was introduced by Sen. Ramon Magsaysay Jr. two years ago, encourages young graduates of agriculture-related courses, as well as budding entrepreneurs who are 21-39 years old to engage in agribusiness and become successful entrepreneurs. The program has initial funding of P50 million (1 million US $) being distributed as grant to participants to augment their operating capital.

Under the program, the government is committed to help participating entrepreneurs to secure loan representing 60 percent of the total capital needed for their proposed agribusiness project. The 25 percent of their needed capital would be provided as grant by the proponent as equity.

"The remarkable development in agriculture, especially the extension of funding support for the implementation of the Agriculture and Fisheries Modernization Act (AFMA) until 2015, would bring about bright prospects to the agriculture sector and make agribusiness a profitable venture, that is why we are encouraging the youth to participate in the YFP," Magsaysay said.

"We are committed to fund projects from a minimum of $935 to a maximum of $5,600 capitalization. More loans and grants would be released soon for approved projects as we need to fully implement the program in order to address the unemployment problem and to assist fresh graduate in putting up their own business", he added.

The One System One Program Approach

The AFMA has basically identified the components of the National Research Development and Extension System in Agriculture and Fisheries. However, there is still an issue of how to reconcile and integrate the Agriculture and Fisheries into one system because they operate by virtue of two different laws.

Essentially, the (RDE) Research Development and Extension System as shown in Figure 2 is described briefly in the Regional Integrated Agricultural Research Centres (RIARCs) operations manual (5) to provide us a framework in unifying the RDE System. The System is anchored on the principle of One System One Program approach where in it aims to consolidate and integrate

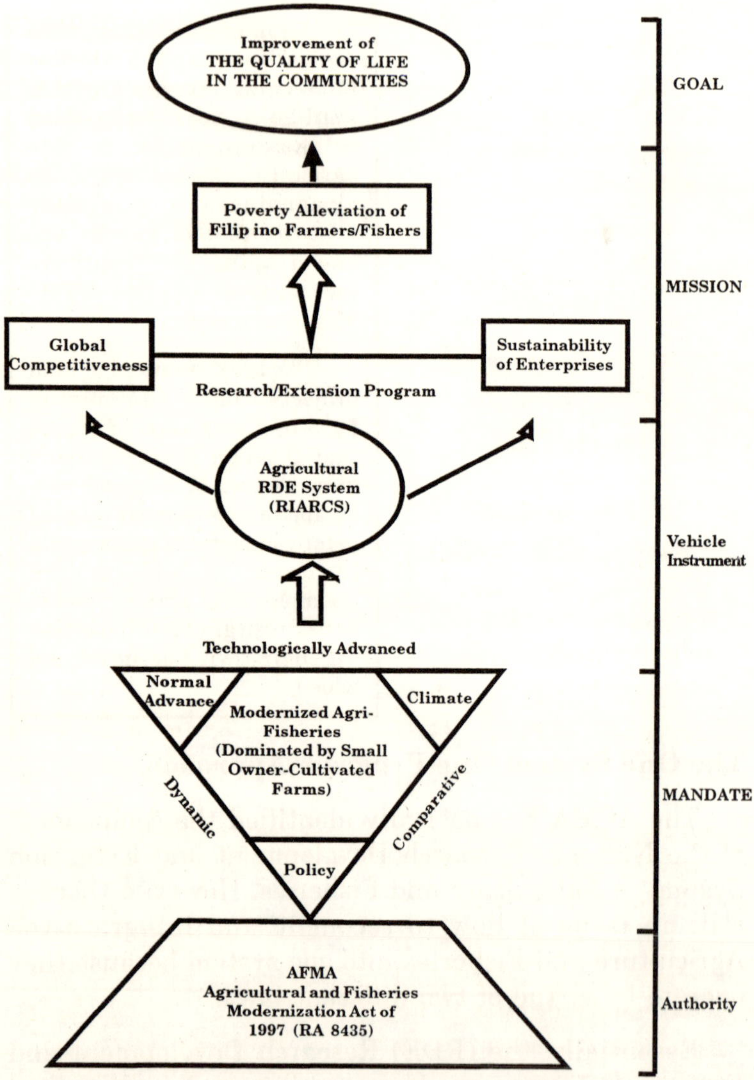

Figure 1. The Ecology and Mandate of Research, Development and Extension

all the research, development and extension program and activities in both the national and regional levels to avoid duplication of efforts and save resources. All R&D agencies, regardless of institutional affiliations and source of funds, are linked on a common agenda and program.

The Bureau of Agricultural Research (BAR) has operationalized this approach by organizing national & regional RDE networks to create synergy among various R&D institutions, guided by a common agenda and programs.

The RDE Convergence

Shown in Figure 2, is a conceptual framework of the National RDE System, a "One System One Program" approach wherein the RIARCs System acts as the convergence point of the technology generation and delivery system. The paradigm shows four entry points as follows:

a. The LGU Extension Research System.

Starting at this first entry point based on the participatory bottoms-up strategy, research must be based on the State of Agriculture (issues and problems) in the locality and its desire for modernity. The assessment coming from the Agri-fishery farm entrepreneurs and consumers is brought up to the RIARCs through the local Government Units (LGUs). The farmers concern can also be channeled through the farmer representative in the (FIACs) Farmer/ Fisherfolk Industires Advisory Committees to the Provincial Technical Institute of Agriculture/Fisheries (PTIA/Fs) and the Agricultural Training Institute (ATI) Farmers Field Schools.

b. Department of Agriculture and other national agencies.

The second entry point in the convergence process comes from the translations of the global and national needs for knowledge and policy directions through the Department of Agriculture and other national government agencies. These policy directions and national agenda are brought down as priority programs to the RIARCs for further integration into regional programs.

c. Specialized commodity-based research centers and private industry.

The need for a specialized knowledge and technologies on national commodities filtered from the global research and national concerns and the technologies generated from private industries are also an input into the RIARCs agenda.

The National Agri-Fisheries Education System (NAFES).

The role of NAFES in knowledge-technology generation/ dissemination and human resource development is a vital entry point to the effectiveness of the RIARCs. It shall focus its research, extension and education efforts to support the effective implementation of AFMA in collaboration with the program priorities in the RIARCs.

The recognition of RIARCs as the venue for the integration of the research, development and extension activities of the DA and the public and private RDE activities in the region is a mechanism of effecting the "one system, one program" principle. It shall serve as the seat of the planning, orchestration of implementation and evaluation of the Regional

Integrated RDE Agenda and program in Agriculture and Fisheries jointly/collaboratively undertaken and/or participated in by various stakeholders.

The RDE Continuum

Figure 3 illustrates the interdependence of research and extension. It shows that the functions belong to a continuum of scientific process that begins and ends with the interest and needs of the farming and fishing communities. In response to the needs of various clients, the RDE continuum also present the types of activities and output generated at different levels of implementation. The arrows show the directional flow and relationships of each sector. It should be noted, however, that in the extension box, the arrows flow towards the farmer client and technology application. It means that in doing extension activities the focus of service is the farm/fisher folks. On the other hand, extension facilitates the application of knowledge and technology produced by research.

The reports of both Contado (6) and Alleyne have described the PLURALISTIC SYSTEM OF AGRICULTURAL EXTENSION in the country. A pluralistic extension system is the contrast of a single, monopolistic agricultural extension service in a given country. It is conceived to address the diverse nature of the country's agriculture and the multiple needs and requirements for assistance of different farmers in different parts of the country. It fosters a range of PARTNERSHIP and utilizes both public and private institutional resources which now also continually gains popularity worldwide.

The system further expands the resource base for the combined process of "facilitator, technology development and technology transfer, especially where

The Philippine Agriculture Research, Development and Extension System

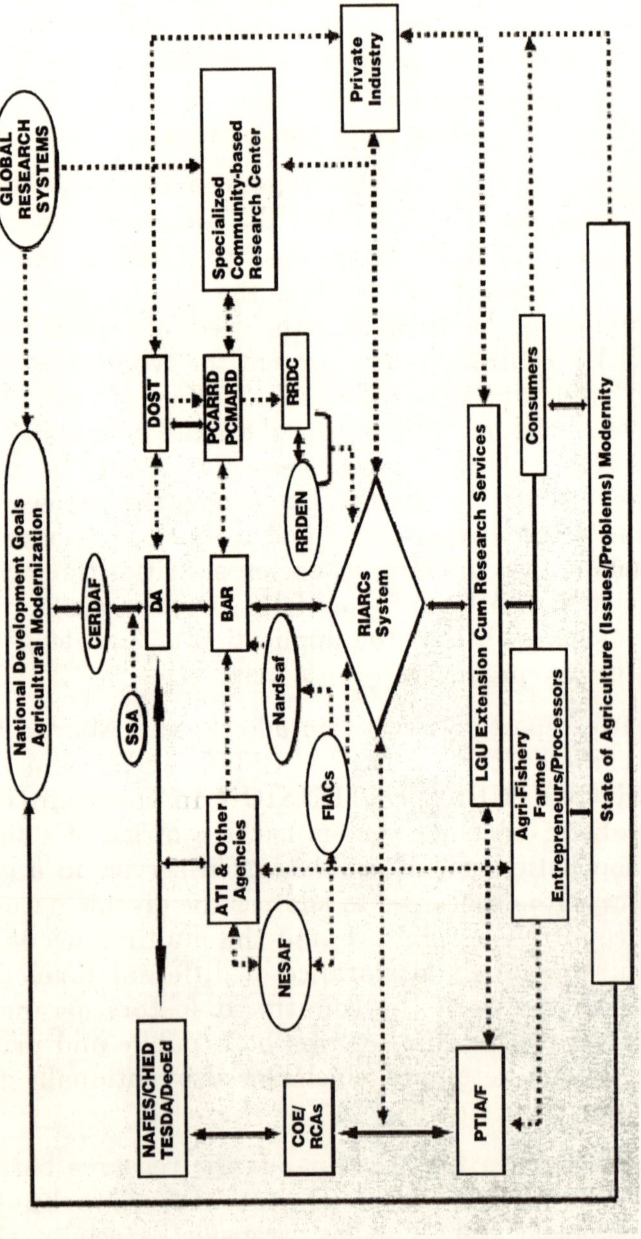

Figure 2. : National Agriculture and Fisheries RDE Convergence System.

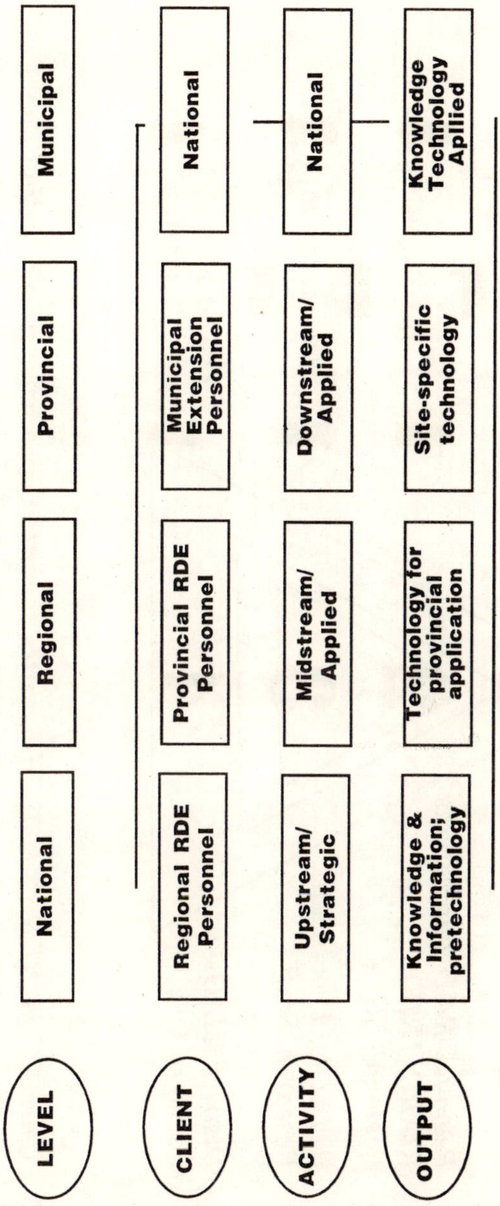

Figure 3. The Research Development Extension (RDE) Continuum and relationships.

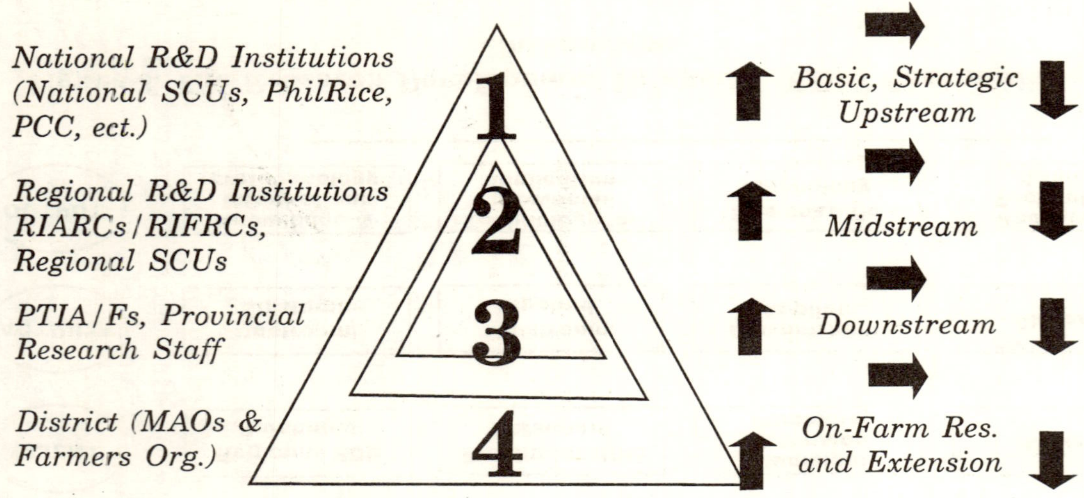

Figure 4. The RDE Pyramid illustrates the interdependence and coordination of the various types of research and institutions and their relationship to extension.

government own resources are limited and possibly shrinking. The true nature of a pluralistic and a decentralized system suggest vigilance, continued coordination and participatory dialogues between the various actors.

Such a pluralistic extension is composed of three subsystems; (a) The national government subsystem which directly compliments; (b) The local government subsystem; and (c) The private sector subsystem. Each subsystem has its own defined roles and functions.

Strategic Planning and Implementation of a National Pluralistic Extension Agenda.

A. The State Policy

Under AFMA, the state policy on extension shall give priority to the utilization of research results through formal and non-formal education, extension and training services. It shall support the development of a national extension system that will help accelerate the transformation of Philippine agriculture and fisheries from a resource-based to a technology-based industry.

The agriculture and fisheries extension services shall cover the following major services to the farming and fishery communities :

a. Training services;
b. Farm or business advisory services;
c. Demonstration services; and
d. Information and communication support services through tri-media.

The delivery of these extension services shall be multidisciplinary and shall involve the farmers, fisher folk, and their organizations, and those engaged in food and non-food production and processing including the private and public sectors.

B. The Over-all Framework of Extension

As defined by the Department of Agriculture Administrative order No. 13 s. 2005, the National Extension Agenda and Programs (NEAP) shall reflect the contributions of the stakeholders of agri-fisheries extension delivery system in the country. It shall address the diverse nature of the country's agriculture and fisheries needs and requirements for assistance at the same time supporting the government's priority commodity programs. It shall also foster a range of partnership in the utilization of both public and non-public institutional resources.

The extension functions of the Department of Agriculture (DA) shall be undertaken by the Regional Field Units (RFUs) and Agricultural Training Institute (ATI) in collaboration with the Local Government Units (LGUs) and State universities and Colleges (SUCs). On the other hand, the ATI Regional Training Center (RTCs), shall DESIGN AND IMPLEMENT PROGRAMS which are CONSISTENT AND FUNCTIONALLY INTEGRATED with the RFU's regional agriculture and fisheries development strategy and program.

C. Extension Agenda and Program Development Process

Based on the above mentioned state policies, the FAO proposed policy framework for Philippine agricultural extension and the National RDE System's one system one program principle, the extension agenda and program can emanate from three levels and converge at the RIARCs System which are described below.

1. At the National level (Top-Down).

The National Extension Agenda and Programs (NEAP) can be drawn from the National Integrated Research, Development and Extension Agenda (NIRDEA), translated into national RDE major program areas (19 major program areas at present under the name National RDE Network). The National Integrated RDE agenda and program (NIRDEP) contains two or more research projects that are complimentary, requiring an interdisciplinary approach to meet established goals within a specific time frame. It details the sequence of activities to address a research or technological problem a holistic, integrated and systematic manner from technology development to technology utilization.

2. At the Regional Level (Convergence Point)

Following the new modality of the "One System One Program" approach, there shall be one Regional Extension Agenda and Program (REAP) for Agriculture and Fisheries that shall serve as the overall umbrella for the DA-RFUs, regional DA attached agencies, SUCs, PTIAFs, private R & D institutions and other agencies mandated to undertake research and extension in Agriculture and Fisheries. The REAP shall cohere with the National RDE Agenda and program priorities or specifically the NEAP. These are translated with major programs at the regional level which are called regional networks. These programs contains distinct but highly complimentary programs of research and extension that are inter-institutional and interdisciplinary in nature. The regional programs address the specific RDE needs of the priority development zones in the region, particularly the identified Strategic Agriculture and Fisheries Development Zones (SAFDZ).

The Regional RDE Network provides leadership in the planning, orchestration of implementation, monitoring and evaluation of the RIRDEP in coordination with the regional training centers in agriculture and fisheries. Its goal is to develop and promote relevant, sustainable and cost-effective technologies to make farming and fishing communities globally competitive.

3. At the LGU Level (Bottoms-up)

While the LGU Extension cum Research System is autonomous by virtue of the local government code, it is a part of the overall RDE Network. The provincial government, through the Provincial Research and Extension Centers and Provincial Training Centers), shall integrate the operations for the agri-fisheries extension services within the province and undertake continuing and periodic annual evaluation of all municipal extension programs. The system shall be multi-disciplinary and shall involve the farmers, fisher folks and their organizations, and those engaged in food and non-food production and processing including private and public sectors and agri-business entrepreneurs.

Starting at this level, the RDE programs must be based on the state of agriculture (issues and problems) in the farming locality and its desire for modernity. These assessments are brought up to the RIARCS, the LGUs, the Farmers/Fisherfolk Industry Advisory Committees (FIACs), the PTIAFs and the ATI Farmers Field Schools.

D. The Thematic Agenda and Programs

One of the functions of the Department of Agriculture through ATI is to provide funding support to extension activities. Section 2.2 of the DA Administrative Order No. 13. S. 2005 provides the following thematic agenda and programs which shall be the focus of funding support to any extension grants in the country:

1. Capacity development of the extension system
2. Capability building of extension workers, farmers/ fishers as managers/entrepreneur.
3. Development of Functional Extension Models.
4. Enrichment/ Improvement of Extension Policies, Mechanisms and Approaches.
5. Monitoring and Evaluation of outcomes and Impacts of Extension Services.
6. Sourcing and Generating Resources for Effective Extension.

E. Pluralism as Enhanced in Cyber Extension

The concept of cyber extension short messaging services as it is being used in India and China is finding its own position in the Philippines.

Recently, the Philippines was cited as "Asia's Leader in Short Message Service (SMS) technology during the CommunicAsia 2004 event held in Singapore (Manila Bulletin, June 27, 2004 p. B-3). The country has an estimated 28 million mobile or cellular phone subscribers and industry estimates that half of the 84 million population will own a handset by the end of the year 2005 (AFP Report 2004). This figure attributed to the affordable and accessible mobile services in the Philippines.

Already dubbed as the texting capital of the world with 120 to 150 million messages crossing mobile phones daily, the country's SMS or texting messages dominate the traffic by a factor of 8 to 1 (text to voice calls) and is expected to accelerate further in the next few years. (Manila Bulletin, June 27, 2004).

Access to information and improved communication are crucial requirements for sustainable agricultural development. Modern communication technologies when applied to conditions in rural areas can help improve communication, increase participation, disseminate information and share knowledge and skills. Cyber extension would now be considered a major form of technology dissemination in the near future (Sharma, 2000).

In a recent study reported by Maningas and Mancebo, 2004, p.3, showed that cellular phones and landline telephones were the most accessible and utilized among the identified communication technology media by the DA and DOST client respondents. In the Cordillera Region it was revealed in an evaluation report that about 80% of households owns cellphones and about 50% of farm households use the cellphones in their communication. On the other hand, the Visayas fishermen and fish dealers already use the cellphones in locating the market with the best price of fish while they are still at sea.

Another study of Mendoza,, 2005, pp. 66-68; 104, in Laguna Province revealed that out of 279 farmer-respondents, 194 (70%) are cellular phones users which they acquired personally while 47 (77.05%) of the 61 extension worker-respondents personally owned and used a cellular phones.

The same study have pointed out that both the farmers and extension workers perceived the acceptability of cellular technology on the basis of its usefulness and ease of use. According to the respondents, the cellular technology was acceptable: (1)

as channel for providing up-to-date information; (2) as a hotline for a wide variety of needs in agricultural services; (3) as a tool for quick information dispatch; (4) for keeping contact of offices and other information services; (5) for texting short messages which extend to wider coverage areas; (6) for timely and wider interaction between and among researchers extension agents, and other farmers and (7) for information dissemination in improving production, marketing and related agricultural activities.

The acceptability of the cellular technology and its effort will undoubtedly empower the small Filipino farmers. Considering further that the Philippines is archipelagic in nature and that communicating to the remote agricultural and business areas is a problem due to distance, the need to optimize the use of cellphones as extension vehicles to reach the sectors' target clientele especially the poor farmers becomes imperative.

It is therefore in this context that the national extension system in the Philippines must realize the changing landscape of information sharing and exchange in the countryside (see Box 3.0) and take note on how the use to the maximum.

F. Strategic Mechanism in Pluralistic Extension

The NEAP as promulgated reflects the true nature of pluralistic extension. From both the demand and supply point of views since no simple agency or organization could adequately meet the numerous extension needs of a varied and large number of farming population, disseminate and transfer information, knowledge and technologies generated by different and numerous public and private research centers, two basic

Box 3.0 Philippine Star September 11, 2005

Mobile technology finds application in farming

With text messaging and mobile computing gaining winder acceptance and popularity in the Philippines, the Open Academy for Philippine Agriculture (OpAPA) plans to use mobile technology to provide information to extension workers and farmers.

The OpAPA or Pinoy Farmers' Internet recently deployed a short messaging system (SMS) contact center through which extension workers and farmers can send their queries and comments.

A Web-based application developed in collaboration with the DOST Advanced Science and Technology Institute (ASTI) allows call center operators to facilitate queries and send broadcast tips on farming technology through text messages.

Incoming and outgoing messages are logged in a database for further analysis and collation, to be included in a growing knowledgebase being established by OpAPA content developers. The system uses the ASTI-developed GSM module in conjuction with the seb application. Another SMS-based service is a seed inventory system where farmers, dealers and growers can transact online. Farmers can check the seed stocks, while dealers and growers can upgrade their seed stock inventory in real time. The project is funded by the Pan Asia Networking International Development Research Center. Plans are also under way to deploy mobile-based Internet connections in pilot and expansion sites of the Open Academy.

"We are also drawing up plans for a mobile Internet bus that will roam the countryside as a 'roadwhow' of the Pinoy Farmers' Internet e-extension services, "Open Academy program director Roger F. Barroga said.

Another future project using mobile technology is the use of multimedia services (MMS), in field diagnosis. Farmers and extension workers can take pictures of their affected crops using cellphone cameras, and send the images to the Farmers' Call Center which will send the query to experts for diagnosis and possible solutions.

"All these are being made possible by current and readily available technologies. What we are doing is simply bringing these technologies within the reach of extension workers and farmers," Barroga said.

"This way, mobile and information technologies become effective and relevant enablers in agricultural development," he added.

principles in pluralistic extension need to be effectively strengthened, Contado (2003). They are as follows:

1. Principle of Partnering

Partnering is nothing more than a strategy that assists people within an organization to work together in a higher effective manner. This involves discussing the common goals and objectives that you with the people with you are entering a partnership arrangement.

2. The Principle of Convergence at the farming communities.

Convergence is one principle that is very vital to a successful extension service. Pluralism means many organizations are brought together as partners in the effort of delivering extension to promote common goals productivity. One important emphasis is that the locus this of convergence is at the farming communities in which the phenomenon of location specificity will be addressed by all the actors in the system.

Conclusion

The long years of Philippine experience in the transformation of agricultural extension and its system of delivery are a good case for study and analysis. Emerging from a single, monopolistic extension system that have been characterized to be chaotic, overlapping and confusing towards a pluralistic system conceived to address the diverse nature of the country's agriculture and fisheries and the multiple needs and requirements of different farmers in different parts of the country, the adoption of pluralism in Philippine extension service under a unified one program approach is a milestone and a good model to watch in accelerating the goals of AFMA in modernizing Philippine Agri-fisheries.

The development and specifications of the details in NEAP led by the ATI can be easily arrived at based on an agreed framework of "market oriented" pluralistic extensions system. While the Philippine model seems complicated composed of many levels and diversity of actors, such will be greatly enhanced by a good leadership, effective partnering and productive convergence of efforts towards increasing productivity, efficiency, income and employment.

Paraphrasing, Dr. B.E. Swanson, in his experience carrying projects in China, India and Egypt, he posited that "although most developing countries now have food surplus, many still comfort the problem of large members of malnourished rural people, especially children. In the process they have begun to recognize that hunger is not a food problem, but a money problem. Therefore, the policy focus has begun to shift from increasing rural income and employment".

The basic proposition therefore according to him is that if extension is going to shift its focus to increasing incomes and employment, then it must shift from being a "SUPPLY-DRIVEN" extension system to one that is more "DEMAND" or "MARKET-DRIVEN".

The effort therefore in pursuing the Philippine model must be towards having more income (farm & non-farm income) and employment among the farming households to attain the dream of a better quality life.

References

AGRICOM Legislative Recommendations Report (1998). Congress of the Philippines.

Agriculture and Fisheries Modernization Act (1997). Implementing Rules and Regulations.

The Dow Chemical Pacific Ltd. and the Family Farm School Experience (2005). The Philippine Agriculture Magazine., May 2005. pp.14-16

Young Farmers Program. A primer. National Agricultural and Fishery Council. Department of Agriculture. Quezon City

The Regional Integrated Agriculture Research Center Operations Manual (2002). Institutions Development Division, Bureau of Agricultural Research, Department of Agriculture.

Contado, Tito E. (2003). Agricultural Extension in the Philippines Status, Problems and Strengths for Effective Technology Promotions. A paper prepared for the Division VI. Agriculture & Forestry Division of the National Research Council of the Philippines on March 7, 2003 at ERDB, College of Forestry Campus UPLB Los Baños, Laguna.

Alleyne, Patrick E. (2002). Towards a Decision on Policy Issues: Agricultural Extension in the context of Devolution: The Philippines. FAO. TCP/PHI/0167. April, 2002.

FAO's Report on the Philippine Agricultural Extension System (2003).

Department of Agriculture Administrative Order No. 13, 5. (2005).

Maningas, R.V. (2003). Information and Communication Technology (ICT) utilization by End-user of Selected Agricultural Research and Extension Networks in the Philippines. Unpublished Ph.D. Thesis, University of the Philippines Los Baños. College, Laguna.

Mancebo S.T. (2004). Agricultural Production System: Household Farm Income and Adoption Rate Survey. The Management of CASCADE Impact Studies (Project ALA/PH/9627) Department of Agriculture. Diliman, Q.C.

Mendoza, Isabel C. (2005). The Acceptability of Short Messaging Service for Cyber-Extension in Selected Farming Communities in Laguna Province, Philippines. Unpublished Ph.D. Thesis, University of the Philippines Los Baños. College, Laguna.

Swanson, B.E. (2004). "Extension Strategies for Poverty Alleviation in a Global Economy" The APEAN Agri-vironment Education Vol.4 No.1 p. 11.

Alan Patching., 2005 Partnering for Productivity. www.newsletter.com

Acronyms

AFMA – Agriculture and Fishery Modernization Act

NAFES – National Agriculture and Fisheries Education System

CSA – Community-Supported Agriculture

YFP – Young Farmers Program

RIARCs – regional Integrated Agricultural Research Center

BAR – Bureau of Agricultural Research

LGUs – Local Government Unit

FIACs – Farmers/Fishfolk Industry Advisory Committees

ATI – Agricultural Training Institute

PTI/Fs – Provincial Technical Institute of Agriculture/ Fisheries

PREC – Provincial Research Extension Center

MLGUs – Municipal Local Government Units

CHED – Commission on Higher Education

TESDA – Technical Education and Skill Development Authority

DepEd – Department of Education

COE/RCAs – Center of Excellence/ Regional College of Agriculture

NESAF – National Extension System for Agriculture and Fisheries

NaRDSAF – National Research Development System in Agriculture and Fisheries

CSA – Community Supported Agriculture/ Corporate Supported Agriculture

DOST – Department of Science and Technology

PCARRD/PMARD – Philippine Council for Agriculture,, Forestry and Natural Resources Research and Development/ Philippine Council for Aquatic and Marine Research Development

RRDEN – Regional Research Development Extension Network

RRDC – Regional Research and Development Consortium

RDE –Research Development and Extension

SCU – State Colleges and Universitics

PhilRice – Philippine Rice Research Institute

MAOs – Municipal Agriculture Officer

NEAP – National Extension Agenda Programs

RIRDEP – Regional Integrated Research Development Extension and Program

16

Agricultural Information Infrastructure Development and Information Services in China

Qijie Gao and Xiaoyun Li

Introduction

At present, China's agriculture is in transition from traditional agriculture to modern agriculture. With the application of information technology in agriculture, agricultural structure will be further upgraded, development patterns will change and competitiveness of agriculture will be strengthened. This paper reviews the situation and achievements of agricultural information infrastructure development and information services in China, analyses the development trends and main problems of agricultural information services, and puts forward some recommendations for developing agricultural information infrastructure and information services in China.

Situation and Achievements

After 10 years of exploration, agricultural information service in China has achieved great success.

In detail, the main achievements in agricultural information infrastructure development are as follows:

First, information service system in agriculture has set up basically. Agricultural information service system ranging from the central government to the local regions has established. Up to now, agricultural departments in 97 percent of the prefectures and 80 percent of the counties around the country have information management and service agencies. There are about 4000 websites concerning agriculture, and 180 thousand government-funded rural information service workers who can transmit information to farmers directly. The experiment of "Integration of TV, Telephone and Computer" has been expanded to 100 counties.

Second, agricultural information network has reached a relatively large scale. Among the 31 provinces, cities and autonomous regions in China, about 80 percent of the prefectural agricultural departments and 40 percent of the county agricultural departments have established local area network, and 80 percent rural information service stations own computers with Internet access available. With years' efforts, an agricultural information service network covering provinces, prefectures, counties and townships has reached initial scale. Based on this network, three applied systems have been gradually improved including rural product quality control system, market supervision and management system, technology and information service system.

Third, integration and exploration of agricultural information resource have been strengthened. The amount of agricultural information resource has been increased, information selecting system has been formed basically, and the coverage of information release becomes larger and larger. Agricultural information resource integration has been reinforced greatly, with its contents enriched. Agricultural information selecting system, focused on production information, market information, technological information and policy information, has formed. The Ministry of Agriculture (MOA) has nearly 40 standing information collecting channels, 8000 information collecting points in the fields of agriculture, animal husbandry, fishery, farmland reclamation and farm machinery. Meanwhile, more than 50 databases were set up, with information resources covering the basic aspects of agriculture and rural economy, and internal information sharing mechanism was established. Since 2002, the monitoring and precaution work on some agricultural products, such as wheat and corn, has been started, which provides timely analysis and prediction on the demand and supply situation at both domestic and international markets. In 2003, MOA issued "Economic information dissemination calendar" system for releasing information in all areas of life periodically. The network of agricultural information service is now rapidly extending to intermediary organizations, leading companies, wholesale markets, villages, brokers and farmers in farming and breeding. At present, there are 40 thousand agricultural leading companies, 170 thousand rural cooperative intermediary organizations, 950 thousand production and managing farmers with large scale, 2400 thousand rural brokers, all of whom can receive information service from agricultural departments.

Fourth, innovations have been achieved in agricultural information service pattern. Different agricultural information service patterns emerged across the country. The integration of TV, telephone and computer, which is being implemented by MOA, is one of the successful innovations by many rural technicians.

This pattern realizes the integration and visualization of information services, and expands the information services coverage effectively. Therefore it gains much popularity among the farmers. Nowadays, many innovations can be observed relating to the pattern of information services, such as "millions-of-farmers-on-line" project, information's entry into villages and rural companies, electronic wave's entry into households, information bulletin, information selection, agricultural information 110 call service. These innovations can fully take the advantages of different information carriers, satisfy individual needs of the agricultural producers, and play an unsubstitutable role in agricultural information services.

Fifth, electronic administration has performed efficiently. In accelerating information services, with the overall extension of computer network, government affairs, such as regulation, guidance and supervision services, have changed a lot. Applicant unit now can get "once for all" services from the integrated administrative and supervision information system. Some local agricultural departments have achieved the publicity of procedure, process and result in supervision, and the operational efficiency has been improved apparently with the development of electronic administrative affairs.

Needs for and Development Trends of Agricultural Information Services

Needs for agricultural information services

The demand for agricultural information is mainly focused on applied technology of agricultural production, information on the demand and supply in the market and information on agricultural products prices. The difference in demand among different regions is obvious: In the east, farmers pay much attention to both the information of demand and supply in the market and the information concerning national agricultural policies as well as applied information and technology of agricultural production; In the middle-west, information on agricultural technology, agricultural products prices, and information on production materials are much needed. The demand of agricultural modernization for information service can be expressed mainly in three aspects: the information service of agricultural infrastructure, the full automation of agricultural technological operation, and the popularization of information network of agricultural economic management. The demand of rural urbanization for information service is focused on the information service of rural education and training, the information service of rural management and the information service of rural finance and taxation.

At present, the main suppliers in China's market of agricultural information service includes operators such as China Mobile, China Unicom, China Netcom, as well as IT software and hardware providers like HP, Oracle, Huawei 3Com, IBM, Nongxintong Technology, Beijing Zhongnongbosi, to name a few. They shoulder the tasks of meeting the needs of markets with the characteristics of agricultural information service. In the face of the broad markets in rural areas, these characteristics contain not only the individuated equipment designs in these companies, but also some innovative frameworks and systems.

Currently, although there are more than 4000 websites of agricultural information, 3000 agricultural periodicals, hundreds of newspapers concerning agriculture and some broadcasting and TV programs on agriculture, most of the demand for information from 800 million farmers still can't be met.

Case

According to the survey by China Rural Technology Development Center (Chen, 2005), the main providers of agricultural information are research institutes, universities, agricultural administrative departments and agriculture-related enterprises. They provide information mainly on agricultural techniques, new varieties, production materials, agricultural development policies, market prices, research results and education on line through channels like books, periodicals, database, electronic bibliography, tapes and videos. Farmers obtain different kinds of information mainly from channels such as TV and broadcast, newspaper and magazines, neighbors and relatives, extension service, rural libraries, technical VCD, specialized technology associations and training. Only 1.8% of farmers use computer networks.

In 2004, another sample-investigation about "the situation of farmers, market, technological information" was carried out in a MOA Survey including 1600 sample households in 30 provinces. This investigation shows that:

The ways in which to get information has been improved

From the investigation, the integration of TV, telephone and computer became common gradually. Now, farmers with telephones account for 50.5% of the total, and farmers with TV sets account for 91.3% of the total. 1.6% of the farmers have computers, and farmers who have access to Internet account for 1.7% of the total.

In 2003, every 100 farm households subscribed to 27 newspapers and magazines on average, and the expenditure per household is RMB 24 Yuan (about US$ 3). Each household bought about one book on agricultural technology on average, and the expenditure per household was RMB 6.85 Yuan (about US$ 0.85).

In the same year, 38.0% of the farmers were trained by technicians on production techniques, of whom 5.3% should pay for the training. The training fee each household paid was RMB 13 Yuan (about US$ 1.6).

Traditional information sources still play an important role

Purchase of production materials mainly depends on experience and the recommendation of agricultural technicians. When farmers buy production materials, the main factor influencing the behavior is their experience, which accounts for as much as 44.3% among all the choices of information sources. The recommendation of agricultural technicians ranks second, accounting for 17.7%. The third is broadcasting and TV advertisements, accounting for 10.7%. On the contrary, the influence of computer network, contracting system is still very little.

Adoption of new technology mainly depends on neighbors' demonstration and reading books and newspapers. In learning and applying production techniques, though traditional ways still play an important role, modern communication methods become more and more influential. Among the sample farmers, learning production techniques from other farmers around ranked first, accounting for 38.0%; Learning from agricultural technicians ranked second, accounting for 21.8%; The third is learning by reading books and newspapers themselves, accounting for 14.5%. But what must be pointed out is that the percentage of " watching channel 7 agriculture program in central TV station" has increased from 8.5% of previous year to 11.9%. It is obvious that farmers pay more attention to learning advanced production techniques through media. However, other ways, such as contracting system, specialized technology association, information release activity, agricultural broadcasting school etc., can only play a limited role.

Selling agricultural products mainly depends on local market and neighborhood. Decision on selling agricultural products is determined mainly by the prices that farmers get in the market, then by the recommendation of village leaders or neighbors, they account for 35.6% and 29.0% respectively. Getting information from purchasing contract accounts for nearly 9%. Selling at home accounts for 8.1%. Receiving selling information through TV, broadcasting and newspapers accounts for 14.0%. Although the amount of receiving information through supplying-and –selling association and computer network is very small, it has increased apparently compared with the same time of last year.

Face-to-face instruction from agricultural technicians gains much popularity

This investigation revealed that farmers preferred agricultural technicians to teach face to face. Among all the sample farmers, 87.0% thought face-to-face instruction was very good or good, better than communication with farmers nearby.

The communication and demonstration among farmers are still very important for farmers to get information. 86.7% in the sample thought this method was very good or comparably good. The third most popular way of getting information is through TV. Percentage of Farmers who thought television information was very good or comparably good had increased by 3.5 percent compared with previous year. The order of popularity among other methods is as follows: books and newspapers, information release activity, blackboard, advertising, computer network.

Price fluctuation is the most important information for farmers

The fluctuation of market price and grain & cotton purchasing price set by the government attracts most farmers' attention. Farmers who think these two prices are important or very important account for 90.1% and 82.7% respectively. Besides, farmers who consider the prices in international market are important or very important account for 40.8%.

Farmers are relatively familiar and satisfied with the conventional information resources

Of the investigated farmers, 75.2% know the Channel 7 agricultural program in the central television station, 68.7% know "Farmers Daily", 76.9% know the lectures given by rural technicians, more than half of the farmers know "Farmers Digest" and the agricultural broadcasting school.

Concerning current rural information dissemination system, farmers are more satisfied with the Channel 7 agricultural program in the central television station, the lectures of agricultural technicians and "Farmers Daily"?"Farmers Digest". Of the farmers who know the above dissemination methods, 84.9% are satisfied with the Channel 7 agricultural program in the central television station. The satisfactory rate of “Farmer Daily” is 76.4%, and 78.3% of farmers are satisfied with the lectures given by agricultural technicians.

Because the development levels in different areas of China are uneven, many farmers don’t know the existing information resources, especially the computer network and the hot line system for science and technology services, which depends on advanced technology. Therefore, more work is still to be done to provide farmers with more and better agricultural information service.

Development trends

In general, there are three important development trends in the field of agricultural information service in China.

First, the demand for agricultural information service tends to be complex. Different sectors, industries, enterprises, cooperative organizations and farmers have different demand for information. Concerning the kinds of information demand, supply and demand, price, labor transfer have turned into new focus of information demand, except production, science and technology, policy information. Regarding the quality of information demand, the original demand mainly for primitive information, single information, static information has gradually turned into one for processed information, value-added information, integrated information, dynamic information and information analysis and forecast. Concerning the main demanders of information, besides government organizations at different levels, agricultural departments at grassroots level and farmers, various kinds of agricultural enterprises, and intermediary organizations have become the new demanders for agricultural information.

Second, the agricultural information service channels become diversified. Because of the regional difference in social information service and economic strength, the methods for promoting agricultural information service vary. Different regions have explored different agricultural information service channels according to their situations. A new trend of integrated use of multiple channels, ways and patterns is emerging. On the one hand, a good use of the special advantages of modern information technology such as computer network, should be made, on the other hand, the conventional information dissemination methods, such as TV, telephone, broadcasting and simple newspapers, could also be effectively utilized.

Third, agricultural information service is penetrating in all of the subsystems of agriculture. The wide use of information technology in agriculture has made agricultural information service pervade in all links, including agricultural production, processing, transportation, research and education, technology extension, consumption, and so on. This makes agricultural information service no longer an isolated conception away from agricultural production, agricultural science and technology, and transportation system, but one tightly connected to every agricultural sectors and industries. In fact, information services could be developed according to the local development strategy and key missions.

Main problems facing agricultural information infrastructure development & information services

Although great progress has been made in rural information service, the extent of the rural information service is still quite low and can't meet the requirement of the modern agriculture development. The latest report of China National Network Information Center (CNNIC) in 2004 shows that the amount of the people who are involved in the business of agriculture, forestry and animal husbandry and have access to the internet only accounts for 2.3%, which means most people related to agriculture haven't connected with the Internet.

The major problems facing agricultural information service are as follows:

The comprehensive quality of the farmers is relatively low, and their information consciousness is weak

At present, the average educational attainment of rural population in China is 6.79 years, 54.17% of farmers have attained less than 6 years of education. The relatively low comprehensive quality of the farmers is related to the low level of education and low scientific quality, which directly limits the ability of farmers to learn the information technology and Internet knowledge, and information application accordingly.

The information consciousness is the prerequisite for farmers to gather, distribute, analyze and apply the information. Due to the undeveloped economic and social situation in many rural areas, the information consciousness of most farmers is weak except for the farmers in the coastal areas and the areas with frequent trade of products and services.

The cost to gain the information is relatively high

The farmers' income level determines their information application too. According to the source from Chinese Statistical Bureau, the annual average per-capita net income of farmers is RMB 2622 Yuan (330 US $) in 2003, among which income in cash is RMB 1721 Yuan (215 US $). In addition, the small scale and scattered lots of farmland of the farmers restricts the utilization of modern information technology. To gain the information via telephone is convenient, but costly. Information channels like TV, blackboard or leaflet etc. is less costly, but cannot meet the needs of the farmers, which to some extent decrease the demand for information of the farmers.

The infrastructure for agricultural information service is not satisfactory at grassroots levels

Although agricultural information has made great progress at the national and provincial level, the conditions at township and village level are not so satisfactory. Most agricultural departments at county, township and village levels in west area of China have no computers and access to Internet. Most existing software or equipments are old-fashioned. Moreover, the operation cost is high, which also limits the further exploration of information resources and development of information service. In urban areas, there are 39 telephones for every 100 persons, while in the rural areas, only 6 telephones.

Source and quality of information is not satisfactory

Farmers are eager to know the accurate information about the growth situation of agricultural crops, future price trend and risk pre-warning of agricultural production. Most relevant information on the Internet is not standardized, but arranged randomly, which makes the quality and accuracy of the information resource lower. Therefore, it is difficult for farmers to get the needed information, and many farmers have doubt on the reliability, authority and value of the Internet information. At present, there are roughly 4000 agricultural information websites, but most of them contain much general information and less specialized information. Most of them are related to simple information gathering, not to classified information specific to concrete conditions. Shortage of valuable information on agricultural economic management and price is obvious. Therefore, agricultural products and market situations cannot be analyzed very well, which makes agricultural information less valuable and interesting for the farmers. All of these limit the effective application of information, and it is not good for the farmers to explore the Internet resources.

The information resources have not effectively integrated and developed

At present, both agricultural departments and apiculture-related institutions have their own information resources. Information sharing mechanism has not been formed effectively. It is difficult for the government organizations at all levels to collect the comprehensive, systematic and accurate information, and it is not convenient either for farmers and agricultural enterprises to get information consultation service.

The information service agents at grassroots levels are not competent

Most of the information service agents at grassroots levels are not specialized in information technology and management, and their ability to treat and upgrade related information is relatively weak. Shortage of information analysts causes that most information resources cannot be fully developed.

Recommendations for agricultural information infrastructure development & information services in China

Improvement of rural information service depends on the government's support and funding. Meanwhile farmer's information needs and information market should be studied.

Training the information service staff. It is important to have high-quality information service personnel. At present there is a shortage of persons with advanced information knowledge, and the structure of service staff is unsatisfactory either. The training for information service staff could be focused on their capabilities of information collecting, processing, treating, feeding-back, analyzing and forecasting.

Improving the information infrastructure and the service system. Although the rural information networks of China have reached the counties and even the townships in some areas, there is a weak linkage between farmers and the county and township agricultural officers. So the future emphases should be placed on the establishment of information service stations at township levels and training of information service workers at village levels. Full use should be made of different kinds of rural service organizations, agricultural enterprises, brokers of agricultural products, broadcasting schools and village leaders, so that they can cooperate tightly with information service system in information collecting, processing, releasing and so on.

Facilitating the information sharing through allocating and integrating rural information resource effectively. The key works to do are pre-warning of agricultural products, supervising of market and information service of science and technology. At present, the related rural information is scattered in different departments, which limits the use of information resource and government regulation. Mechanism of information communication and sharing is to be improved so that government organizations, enterprises and farmers can get full and reliable information.

Developing diversified terminals to provide farmers with multiform and individual information service. The government could provide guidance for related equipment manufacturers to produce more practical and cheaper machines and parts. Farmers' information needs in different regions and categories have great difference, therefore service models should be improved to meet the multiplex need of all kinds of farmers.

Choosing suitable ways for farm households to access to information service. According to the experience of some provinces, the alternative ways could be as follows: (1) Sending Agricultural TV Programs to households. Agricultural bureaus at county level set up rural information centers, collect and analyze related information, make TV programs and broadcast them. (2) Special hot linc. Agricultural bureaus at county level build databank and answer farmers' question by telephone automatically. (3) Short message by cellphone. (4) Online service. In some areas, the development tendency is the integration of TV, telephone and computer.

Providing farmers with systematic training. Through training, farmers could have: (1) stronger technology consciousness, which help them to learn and to use the technology to deal with the problems they face; (2) stronger market consciousness, which help them to learn the knowledge of market and to do things according to the market analysis; (3) stronger information consciousness, which help them to be willing to pay for the information they get and give them the consciousness and ability of utilizing information resources.

Fostering farmers' demand for information service. It is important to create farmers' information demand and to inspire them to participate in agricultural information service so that interaction could be achieved between government, service providers and farmers. Fostering and creating farmers' information demand needs to analyze farmers' information needs in their whole production and consumption process. Information about agro-meteorological forecasting, farming, insect & pest, fertilizer, irrigation, biological technology, market and laws and so on is all what the farmers want to learn. Analysis on risk, cost-benefit and market should also be offered for farmers to avoid relevant risks. After the harvest, the products will be stored, processed and packed. Therefore farmers need to learn information on machines, processing, packaging, transportation and food safety.

Improving the management system of rural information service. Government organizations should play a major role in agricultural information service and promote the socialization of rural information service. As public goods, agricultural information service provided by government organizations should face rural extension agencies, agricultural websites, leading agricultural companies, rural information service corporations, and trade organizations of agricultural products. Free information service as public goods should be offered to farmers, and the existing rural extension system could be the main channel of offering public information service. On the basis of function division, cooperation and sharing mechanism should be bettered to avoid rebuilding and to improve the operation efficiency of information system.

Conclusions

Since 1990s agricultural information infrastructure development in China has achieved great success especially in database?information network and service system. The main challenges facing agricultural information service are shortage of agricultural ICT talents and small scale and scattered network stations, which cannot provide standardized service and effective information sharing. Establishment and improvement of agricultural information infrastructure and network service is necessary for the extension of agricultural technology and the integrated development of rural China. Therefore, more attention should be paid to the training of agricultural ICT service workers and coordination of different kinds of agricultural information service agencies and their works, as well as the improvement of existing agricultural information network and service system.

References

Chen, L. (2005). On the Development of Rural Information Service. *Paper presented at the National Conference on Rural Information Service,* Zhuhai, China.

Niu, Y. & Zhou, Y. (2005). Agricultural Information Service in China: Development and Prospects. *Theory and Practice of Library*, 2005(3).

Wu, Z. (2004). Rural Information Service in Development. *Farmers' Daily, www.agri.gov.cn.*

Zhang, Y. (2005). Development of Agricultural Information Service in China. *Farmers' Daily, www.agri.gov.cn.*

17

ICT and Agricultural Diversity

Sangamitra Ramachander and Ashok Jhunjhunwala

Introduction

Experience over several decades has shown that no single approach to agricultural practices produces the optimum output from land everywhere. Plot-to-plot variation in soil composition, micronutrients, moisture availability exists and scientific inputs vary depending on the specific crop and the stage of crop growth. Post-harvest techniques, which add value to the produce, are not common for all crops. Therefore continuous interaction between the farmer and the agricultural scientists is essential to determine relevant and appropriate action. Such interaction is only possible if farmers have access to the scientists and agronomists.

Furthermore, in a scenario of multiple and integrated markets with frequent price fluctuations, the farmer must know where and when to sell in order to maximize profits. To mitigate the unpredictability involved in agriculture arising from market changes as well as inclement weather, water scarcity, crop disease, pest attacks and so on, a farmer requires an appropriate form of crop insurance. He can also want to reduce market risks by trading in derivative instruments such as Options and Futures. Clearly there is a need for the farmer to be connected at every stage to a suitable knowledge and information base, and a source of appropriate training and advice. The role of Information and Communication Technologies (ICT) therefore becomes paramount.

In the era of a more generalized treatment of agriculture, a one-way system of information dissemination (radio, television or printed brochures) was used by agricultural experts to communicate with farmers of a particular region. Today a system is needed that can support interactive communication. The farmers need to interact with experts located elsewhere to discuss their specific situation and at the same time the scientists have much to learn from the farmer. Such learning pertains not only to the specific characteristics of the plot of land, but also to the traditional practices that have been carried on by the farmers, with their innate wisdom of sustainability and ecological balance. An approach that combines both traditional and modern practices in agriculture to optimise the farmer's profitability in a sustainable manner can therefore come about.

In this paper we look at some of the potential areas in which ICT can strengthen Agricultural Extension broadly, and specifically, within the Indian context. This is by no means an exhaustive list, but merely an attempt to demonstrate the nearly limitless ways in which ICT can be used to improve the performance of this sector.

The Importance of Agriculture

Agriculture is an important sector of any economy because it feeds the population and therefore carries a certain strategic importance. The sector also provides the raw materials required to drive some of the key manufacturing industries. A fall in the relative share of agriculture (as a percentage of GDP), compared to manufacturing and services, is considered a normal pattern of development, but in absolute terms the output from this sector is not expected to decline. Therefore it remains important in the case of developing and developed countries alike. Today, agriculture includes high-value areas such as horticulture, floriculture, aquaculture, and livestock rearing. In many countries these sub-sectors are important for export and also have a large internal demand. To meet the needs of the high value agricultural products, downstream industries for agro-processing have developed and so also newer forms of transport, storage and retail. Sophisticated post-harvest technology is used for refrigerated and bulk transport, storage and warehousing, grading of produce and helps in meeting phyto-sanitary standards in the world markets. Therefore the sector has come a long way in terms of modernization, diversification of products and expansion of geographical markets.

The Importance of ICT in Agricultural Extension

Information and Communication Technologies (ICT) have an important role to play in agriculture. A definition from the Overseas Development Institute (ODI) report on ICTs and Rural Development is as follows; "ICT are those technologies that can be used to interlink information technology devices such as personal computers (PCs) with communication technologies such as telephones and their telecommunication networks. The PC and laptop with e-mail and Internet provides the best example". The report goes on to say that the range of technologies that can be included within ICT is increasing all the time and that several other devices such as digital cameras, digital video cameras and players, personal digital assistants, slide projectors and mobile telephones are also becoming compatible with these technologies. Therefore "an expanding assembly of technologies that can be used to collect, store and share information between people using multiple devices and multiple media" can now be categorized as ICT. While the term ICT is of fairly recent origin, there have been other forms of communication used in the past, such as radio, television and printed documents, which have now been termed "traditional ICT". Their main distinction from "modern ICT" is the fact that they permitted only a one-way system of information dissemination, while the latter allows for a dialogue. This difference has vast implications for the use and relevance of these technologies. In this section we explore the role that ICT can play in strengthening agricultural extension, and its particular importance in today's world.

Over the past three decades in particular, much has been learnt about how to improve productivity in agriculture. It is now being increasingly recognized that 'monoculture' is damaging to the balanced development of agriculture and that diversity is at the heart of any biological process. The growth of crops is a complex process and depends on a range of factors that include the topography of the land, the physical and chemical composition of the soil, the moisture content of the soil, the amount of rainfall, the type of seeds, pesticides and fertilizers used, and so on. There is a significant amount of variation from plot to plot in these aspects, and therefore specific measures are required in order to optimise production.

In the past, the farmer used traditional knowledge and practices to optimise output and was an expert in his own right, as he knew the systems well and also had an intimate knowledge of his own land. However, with the advent of modern practices in agriculture, the farmer faces a new range of inputs and must learn the methods of appropriate use and possible consequences. Information and Communication Technologies (ICT) can help impart this knowledge and training to farmers and through a process of interaction between agricultural scientists and farmers, the optimal combination of traditional and modern inputs required can be determined.

In an increasingly integrated world-economy where there is high degree of price volatility, it is important that the farmer has access to information that will help him understand future price trends so that he makes the right decisions regarding what to produce. He also requires price information in order to decide where to sell his produce in the context of multiple markets and varying prices. The need for credit and insurance is also more important today than ever before. ICT can act as an enabler in all of these areas.

Profitability in agriculture is determined both by supply side factors relating to optimising production as well as demand side factors pertaining to the market. We first look at how plot-specific solutions in various aspects of agriculture can help improve productivity and profits on the supply side.

Cropping Patterns

Healthy soil is an essential requirement for good crop yields of high quality. Application of manure and planting of high residue-returning crops have a positive impact on soil health, while excessive ploughing and poor residue-returning crops tend to reduce the health of the soil. Traditional practices recognize the need to maintain soil health by choosing appropriate crops and using a balanced pattern of cropping. However, in today's market economy the farmer must grow the crop with the maximum demand in order to make a profit, and therefore he is unable to incorporate sustainable practices as before. Intensive cropping, characterised by short crop cycles, makes the soil vulnerable to pest attacks. Over time the quality of the soil is reduced, and crop yields and quality decline. Through interaction with experts, for the farmer can determine a more optimal cropping pattern, which would allow him to strike a balance between short term profits and long term sustainability (see, Sanchez and Swaminathan, 2005; where they discuss the need to integrate crop and animal production).

Modern Inputs

As hybrid seeds, chemical fertilizers and pesticides replace their natural counterparts in agriculture, the traditional knowledge systems in the villages are unable to cope. Modern inputs can change the number of cropping cycles that are possible in a given period and increase productivity. Mechanization can also free up time for the farmers, allowing them to diversify their incomes either within agriculture or outside, and reduce risks. However, the appropriate use of these modern inputs requires the traditional knowledge of the farmer to be enhanced and supplemented.

Choosing the right kind of chemical fertilizer for a particular area is important because soils vary in their nutrient supplying capacity and may be inherently rich or deficient in different nutrients. Moreover, nutrients are present in different soil layers, and the fertilizer mix should provide those nutrients that are easily accessible to the root system of the crop at its present state. The need for nutrients also varies by type of crops, with High Yielding Varieties (HYVs) requiring higher NPK application, for example. Through dialogue between farmers and experts, a suitable combination of chemical and natural fertilizers can be determined. Their relative benefits can be weighed from the point of view of cost to the farmer, sustainability and also the health implications to the final consumer. In the absence of proper guidance, farmers may make ill-informed decisions or be misled by the aggressive marketing efforts of competing companies that attempt to promote their specific brand of chemical fertilizer or pesticide.

With infrastructure for communication in place, it is also possible to devise systems of shared use of expensive inputs such as agricultural machinery, which may be unaffordable to most farmers on an individual basis. These can be collectively purchased by a group of farmers from neighboring villages.

Water Use

Water resource analysis, planning and management can help achieve a much higher level of efficiency of water-use in a particular area and save large quantities for future consumption. Water conservation or diversion from other sources based on an understanding of the precise topography of the land and the location of water bodies in the village can prevent a situation of scarcity. Also, distribution and delivery losses can be significantly reduced. Over-use, which leads to depletion of the water table and soil salinity, can therefore be avoided.

An example of a local traditional solution for irrigation that is both efficient and cost-effective is that of the 'Ery' system used in South India, where no perennial rivers exist. Farmlands were converted into areas of temporary catchments during rainfall. The water was then redirected into the nearest water body (tanks) or was soaked into the soil. Therefore conservation took place either through storage or through increasing the level of the water table. This was one of the finest systems of irrigation, which has been uprooted in many areas to be replaced by modern methods. Unfortunately the latter has been deployed on a mass scale without understanding the precise nature of the landscape and water bodies in a particular area. The net result is that our systems of water use, storage and conservation have seen a decline over time. ICT can help in efforts to revive such traditional systems through consultation with those who live in the villages and who best know the topography of the land and the historical systems that were used. Newer methods can be suggested where appropriate to strengthen these.

Credit

Modern agriculture is becoming increasingly commercial and the input requirements of the farmer are therefore much higher than before. Traditional borrowing sources have dried up and the formal sector for rural credit today has an uneven penetration and is of poor quality. Cooperative banks, commercial banks, micro finance institutions, non profit institutions, the government, and non banking financial corporations all exist notionally, but in reality access to credit still remains a huge obstacle in the lives of most of the rural poor. In order to enable the farmer to invest, it is imperative that credit is available on a large scale and ICT may turn out to be an effective means towards this.

In order to set up a well-functioning and sustainable credit framework, there is a need for detailed information on each farmer that would allow the lending agency to assess his or her credit worthiness. Information on personal health and education, family size, assets owned, other loan obligations, scale of farming, other businesses owned etc is factored in when making such assessments. Unfortunately, a database of this kind is lacking in the rural context. An Internet Kiosk in a village can be used to create a database of information, which is an essential pre-requisite for rural credit to be disseminated on a large scale.

These are some of the important ways in which ICT can be used to improve productivity and profits in agriculture. Profit maximization also includes other key aspects on the demand side such as are what to produce, how much to produce and for whom to produce. Unfortunately due to information constraints farmers are unable to make optimal decisions in these areas. In the following sub-sections we look at how access to information and communication can improve profitability and efficiency on the demand side.

Future market prices and risk mitigation

Farmers lack information on expected future market prices during the sowing season and therefore their cropping decisions tend to be a gamble. Moreover, with little knowledge of the cropping decisions of other farmers outside his village, the farmer is ignorant as to whether his choice of volumes is appropriate. Over-production produces a glut in the market and brings down the prices while the choice of the wrong crop could wipe out the farmers' incomes. Given the impoverished condition of the majority of farmers, they tend to be very risk-averse in the face of so much uncertainty. The net result is that investment suffers, and so also overall growth of the sector. If safety nets were in place that could mitigate the risks of production and prices, the farmer would be far better off.

One of the ways to mitigate risks is by trading in Options and Futures. A 'Future' is a paper promissory note that requires the holder to buy or sell a commodity at a particular price at a future date while an Option is a similar instrument that carries the right to buy or sell but with no obligation. Since the price is pre-determined, it is known during the sowing season, and therefore the farmer can make bolder investment decisions. By hedging through the Options market the farmer can lose at most the amount paid as Option premium, whereas the profit potential is unlimited. On the other hand holding a Futures position limits profits, since the farmer is obliged to sell at a particular price, but it also limits the possibility of losses by the hedged price. Telephone and Internet Connectivity in a village can give farmers the ability to access market prices, understand potential future demand trends, and transact in Options and Futures to mitigate market risks. The farmer could also mitigate risk from crop failure through crop insurance, which would help him guard against certain risks during production such as pest-attack, crop disease or natural disasters. As with credit, sophisticated forms of insurance can only exist where detailed profiling of farmers is possible, with both personal and professional background information, and here too ICT can be of help.

Information asymmetry arises when the farm produce is ready to be marketed, as the farmer does not know the prices prevailing in alternate markets. His choice of where to sell is largely a guessing game. The middlemen, on the other hand, are able to benefit from the arbitrage opportunity arising out of the farmers' lack of market information. Timely information pertaining to the prices of agricultural products in various markets would lead to greater market efficiency and more profits for the farmer. Opportunities for arbitrage would also be reduced and the consumer would enjoy a market-determined price. In more diversified agriculture, or high-value agriculture, market cycles are shorter than those for traditional crops and coordination between the farmer and buyer is all the more important. ICT is also necessary as the scale of agricultural operations grows, in integrating the supply chain over wider geographies and improving the efficiency of business processes.

In today's increasingly inter-dependent market, another important aspect is the meeting of international quality standards. This necessitates proper handling of produce, and standardized grading, weighing, packaging and storage. Besides advanced post-harvest technology, information and communication is necessary at every step in order to ensure proper adherence to procedure, transport coordination and timely delivery.

Having established the critical role that information and communication plays in strengthening agricultural extension, the question that follows is, can we provide such infrastructure in rural areas? The following section will answer this question in the Indian context.

Information and Communication Infrastructure

Providing Internet connectivity to 630,000 villages in the country is by no means a small or easy task. One of the early efforts in this regard in the Indian context was to take optical fibre to almost all taluka (county) headquarters and even the smallest towns. This was led by the Department of Telecommunications, and later by Bharat Sanchar Nigam Limited (BSNL), the state-owned incumbent, and has made it possible to provide the highest bandwidth to these areas. Over the last few years several new telecom operators including Reliance, VSNL, Bharati and Railtel have also taken fibres to many talukas. The competition has enabled an inexpensive rural backbone to be set up and Rural Service Provider companies such as n-Logue have created this rural backbone by leasing bandwidth from the telecom operators Such a rural backbone is an essential forerunner to connecting every village.

Equally fortunate is the fact that wireless technology over the last decade and half has made great strides, with the bit-rate (throughput) going up and prices declining continuously. Wireless extensions with a radius of about 15 km taken from the towns that have the fibre could cover nearly 85% of Indian villages and therefore it is now possible to bring Internet connectivity to a majority of the villages in the country.

Wireless Connectivity

Given that connectivity to villages is now possible, the questions are what kind of connectivity and at what cost? So far, the wireless systems available have been GSM (Global System for Mobile Communication), CDMA (Code Division Multiple Access) and corDECT (Wireless in Local Loop: CorDECT is a brand Wireless in Local Loop Product of Midas Inc. & lt; /DIV><). With their latest versions like GPRS (General Pocket Radio Service) and 3G-1X (A standard for a version of 3rd generation mol product), these systems can provide close to 30 kbps dedicated-connectivity to every village. CorDECT can provide up to 64 kbps. While a 64 kbps dedicated connectivity to a village is a good start, the applications needed to empower rural communities (involving video, for example) may require greater bandwidth. Fortunately, emerging wireless technologies appear to be meeting this challenge. 3 G mobile technologies, Broadband corDECT, WiMAX (World Wide Interoperability for Microwave Access), Flash-OFDM (Orthogonal Frequency Division Multiplexing) and iBurst (a product brand name of broadband) are some of the potential options in this regard.

The third generation mobile systems, an outgrowth of the current GSM and CDMA technologies, that are starting to appear are more spectrum-efficient and will replace the current mobiles in the coming years. These 3G systems are indeed capable of providing Internet connectivity, but the absence of a dedicated Internet connection at a bit-rate of over 64 kbps remains the key challenge in the rural context. The major advantage of these systems is their low cost, since they make use of the existing infrastructure commissioned to provide mobile voice services.

The Indian Institute of Technology, Madras, and Midas Communications have continued to innovate using DECT (Digitally Enhanced Cordless Telecommunication) technologies to come up with a Broadband version of the CorDECT (a new technology) system. These new systems provide spectrum efficiency more than five times that of 3G systems and two and half times that of newer technologies such as WiMAX (in its most developed form). Broadband corDECT can provide dedicated 256 kbps to each village today and if sufficient spectrum is allocated, it could even provide 512 kbps connections. The systems are also extremely cost-effective.

As CDMA and TDMA (Time Division Multiple Access) technologies reach their physical limits to provide higher bit-rates, a new technique in the form of

Orthogonal Frequency Division Multiplexing Access (OFDMA) or WiMAX has emerged, which promises to take us forward. These systems could pose a challenge to 3G Mobile technologies when they become cost effective, but this is unlikely to happen before 2008 or even 2009. Once WiMAX is ready and becomes cost-effective it would be of great benefit to rural areas, as it would have mobility as well as support high bit-rate.

There are a few other proprietary technologies that could be very useful today if their cost can be brought down such that they are affordable to rural areas. Flash-OFDM technology, which uses OFDMA, is fully developed and performs today in a manner similar to what WiMAX promises two years hence. Similarly, iBurst is a system that uses tomorrow's multi-antenna technologies to provide connectivity similar to what is promised by WiMAX. Both these systems could provide 256 kbps or even 512 kbps connectivity to each village.

Remote Rural Areas

Despite these considerable advancements, approximately 15 percent of the villages in India are in remote areas where the presence of fibre is low. These tend to be the hilly, forested or desert areas where the population is sparse and the average incomes are lower than in the plains. Laying fibre in such areas is both practically difficult and very expensive. In a number of places, Point-to-Point Microwave Links have been used to provide backbone connectivity, and it is possible to upgrade these microwave links to provide bit-rates as high as 155 Mbps when the cost of the equipment comes down. The annual spectrum charges for the microwave links, however, would be a dampener. Wherever such microwave links exist, it is possible to use terrestrial wireless in about a 20 Km radius of the end points of the microwave links. However, even microwave links are not always prevalent in these areas, since the terrain and vegetation may render it difficult to put such links in place. Under these circumstances, satellite connectivity is a possible alternative. However, providing a satellite connection to every village tends to be prohibitively expensive, especially if a reasonable sustained bit-rate (e.g. 33 Kbps each way) is required.

Sparse Area Communication System (SACS), which combines satellite and terrestrial wireless systems, is a potential option. In this system a satellite remote terminal with a 128 or 256 Kbps dedicated connection (both ways) is installed in an elevated area and a terrestrial wireless system provides connectivity from this remote terminal to about 50 to 70 villages within a radius of 15 to 20 kms. Since the satellite remote 50 to 70 villages share terminal and recurring costs associated with the satellite segment, the cost is not as high. Such a Sparse Area Communication System (SACS) is being developed at IIT Madras along with the Indian Space Research Organization (ISRO) and is expected to be ready in a few months.

The combination of terrestrial wireless with microwave links and SACS systems may serve most of the 15 percent of the villages that are not served by traditional means. However, 1% of Indian villages may still be located too far from each other in which case they cannot be covered by any of these methods. In these cases Direct Satellite Communication to the villages may be the only solution.

Thus technologies are rapidly evolving to make data and voice exchange in real time possible and affordable. With such ICT infrastructure, it is possible to have a more enlightened and balanced approach to agriculture.

However, equally important for the sustained functioning of such ICT in the villages are the kinds of organizational and business models that support it. One example of such a business model is that of the Indian company n-Logue Communications Private Limited.

A Business Model for ICT

n-Logue Communications Private Limited is a Rural Service Provider incubated by the Telecommunications and Computer Networks (TeNeT) group at the Indian Institute of Technology, Madras. The company's mission is "to significantly enhance the quality of life of every rural Indian" by setting up a network of wirelessly connected Internet kiosks in villages throughout India using a for-profit business model that is both scalable and financially sustainable. n-Logue uses a fibre backbone and corDECT Wireless in local Loop as the access technology for its IP (Internet Protocol) network.

The n-Logue three-tier franchise business model consists of the company at the top, Local Services Providers (LSPs) in the middle and the kiosk owners (KOs) at the bottom. The model involves creating local entrepreneurs who own the kiosks and drive the business at the village-level. The company helps the KO set up a kiosk by providing her or him with the necessary hardware, which consists of a Multimedia PC, a digital camera, printer, and the accessories required for providing connectivity. The total cost of the "kiosk kit", including training and six months unlimited connectivity is nearly US $ 1000. The KO needs to earn close to US $ 10 per month to break even and make his monthly instalment payments for his investment.

Once the kiosk is in operation, the KO can access a host of services developed by n-Logue. These include computer education, adult literacy programs, agriculture-related services, health services and government services. The role of the LSP in this model is to set up the Access Centre jointly with n-Logue in order to provide last-mile connectivity to subscribers within his designated project area. He or she is responsible for managing the kiosks in a particular project region, and his responsibilities include recruiting KOs to invest in and set up kiosks in their respective villages, maintaining the Internet and telephone connections, providing technical support as required to KOs, and collecting telephone and Internet fees. As n-Logue's kiosk network grows – there are 2000 kiosks deployed today - the dream is to connect every village in India with Internet and telephone access.

Having examined the feasibility of providing Internet connectivity in a village, the following section looks at the experiences in using such Internet kiosks in providing services.

Examples of ICT in Extension

Despite the very encouraging progress in terms of technology in rural areas and the tremendous potential for its use in agriculture, in reality very little has been done to use this efficiently and on a large scale. The best effort, by far is that of ITC, a company that procures agricultural commodities at a large scale. It has focussed entirely on agriculture and has used the Internet connections in the villages to improve its procurement.

ITC's e-chaupal project covers over 8,000 villages in Madhya Pradesh, Uttar Pradesh, Maharashtra, Karnataka, and Andhra Pradesh providing farming-related information to the agricultural community in these regions. Farmers can look at weather forecasts, order fertilizer and pesticide, and consult agronomists by email. By helping the farmer identify and inputs and farming practices, and by offering a better price for high quality products, ITC is able to ensure that quality standards are adhered to. This helps create a definite competitive advantage for the company.

n-Logue Communications provides agricultural services through its network of over 2000 rural kiosks in the country, by partnering with solution providers. Experts at the Tamil Nadu Agricultural College and Research Institute provide advice on improved farming methods and solutions to crop and animal diseases via videoconferencing. Through the kiosk, farmers can also access various agricultural portals to find weather information, crop prices, and improved farming techniques. Some of these portals include JFarmindia.com, kvbaramati.com, krishiworld.com, and agriwatch.com. Other agricultural services offered by n-Logue through the kiosks are the sales of tractors and spare parts in partnership with farm equipment manufacturing companies. Also, in areas where livestock rearing is a common occupation, kiosks provide online veterinary advice. Potential future agricultural services planned by the company include facilities for soil testing at the kiosk, online farming "schools", an early warning system for disease outbreaks, and long-term predictions on crop prices, demand, and rainfall.

There are several other smaller efforts, which have focussed on building portals and providing services. ISAP, (the Indian Society of Agri-business Professionals), is a not-for-profit organization that aims to create a consortium of experts in the field of agribusiness and rural development. The goals of ISAP are to create a database of both scientific knowledge as well as traditional practices in agriculture to increase productivity, and also help farmers tap domestic and global markets better by finding solutions to the various bottlenecks in the sector. The front-end service delivery points are kiosks of n-Logue, Drishtee, Tarahaat, and other partners include One World and Agriwatch. The services of ISAP are in the form of providing advice on technological inputs and crop diversification, answering queries by phone, email or post, and customised market research and training.

AAqua, an acronym for Almost All Questions Answered, is an online multimedia question and answer forum for agriculture started by the Media Lab Asia centre at IIT-Bombay. The site, apart from providing a platform for Q&A with experts, also allows members to create, view and manage their own content in the local language. In the current deployment farmers have to approach kiosks to send their questions, and kiosk operators charge farmers Rs 10 per query for the consultancy. Among the future plans are for aAqua are to integrate with a cellular phone network for easier delivery of service to farmers.

The Directorate of Marketing and Inspection of the Government of India, in the Ninth Five Year Plan, launched an ICT project called Agricultural Marketing and Information System Network (AGMARKNET), to link all the important Agricultural Produce Wholesale Market Committees (APMCs), State Agricultural Marketing Boards/Directorates and DMI regional offices throughout the country. The basic purpose of AGMARKNET is to ensure effective information exchange on market prices, so that the farmer has the widest possible information on prices, thereby enabling him to make the best decision as to where to sell his produce.

While these efforts are a good start, the emphasis in the future should be on increasing local relevance and the scale of operations.

A Concluding Note

In summary, the key to optimizing agricultural output and creating wealth for the farmer is optimizing the practices of agriculture for each plot. It is based on the oft-forgotten truth that only the farmer who knows his plot best and cares the most to optimize his output. But as agriculture gets modernized with modern seeds, chemical fertilizers and pesticides, the traditional knowledge system proves inadequate. Further, in the era of globalization the farmer does not deal with the local market alone, of which he has some reasonable familiarity. Today even distant events can significantly affect the market price of his produce. ICT becomes extremely important here as it provides a means for the farmer to gain new knowledge and interact with participants in this market. Eventually, however, the farmer will benefit only if his knowledge system is able to incorporate the new inputs he gains. The Internet kiosk in the village could play an extremely important role in bridging this gap.

The kiosk should become a village knowledge center, through which a farmer can access all kinds of information and interact with experts, financial institutions and the market. A "farmers' club" can be set up in every village kiosk that meets at least once a week. The farmers could exchange information and knowledge amongst each other and use the Internet (through the kiosk operator) to access information and interact with farmers in other villages, thereby enriching their knowledge base. Further, farmers could also have access to experts in research institutes, ask questions and discuss their particular problems. Such interaction would enable farmers to cope better with new technology and use it appropriately in the context of their particular plot. The Internet kiosk would also enable them to better understand the global market and trends in commodity pricing. They could learn how to bid in Futures and Options. Finally, the Internet can give them access to financial institutions to help meet their credit and insurance needs. The Internet kiosk in the village therefore holds the potential to become not just a communication center but also a knowledge center for the village.

References

1. www.aaqua.org
2. agmarknet.nic.in
3. www.isapindia.org
4. www.itcportal.com
5. www.n-logue.com

Chapman Robert and Slaymaker Tom, (2002). ICTs and Rural Development: Review of the Literature, Current Interventions and Opportunities for Action, *Overseas Development Institute (Odi), Working Paper 192, London*.

Chowdhury Nuimuddin, (2001). Information and Communication Technologies in 2020 Vision Focus 7: *Appropriate Technology for Sustainable Food Security,* edited by Pinstrup-Anderson Per, International Food Policy Research Institute, Washington DC.

Jhunjhunwala Ashok and Ramachander Sangamitra (2005). Commentary: The Role of Wireless Technologies in Connecting Rural India, *Indian Journal of Radio and Space Physics, Vol. 34.*

Jhunjhunwala Ashok and Ramachander Sangamitra, Rural Connectivity in India, *CSI Communications,* August 2005.

Richardson Don, *How Can Agricultural Extension Best Harness ICTs to Improve Rural Livelihoods in Developing Countries* Guelph, Ontario, Canada.

Sanchez P., M.S. Swaminathan et al. (2005). *Halving hunger: it can be done. UN Millennium Report Task Force on Hunger.* London, Earthscan.

18

Conclusion

Changes Needed in Extension Programmes and Extension Research Needed to Support these Changes

A.W. van den Ban and R.K. Samanta

Extension programmes

This book tries to help managers of agricultural extension organisations and extension scientists in Asia to decide which changes in the roles of the agricultural extension organisations in their country are desirable and how these changes can be implemented. We can not make recommendations about these changes, because which roles are desirable and possible depends on the local situation, the competencies of the extension staff and the policy goals of the organisations which promote and finance extension. In this chapter we will mention changes which in our opinion should be considered. Therefore we will highlight several of the changes mentioned in previous chapters.

Mancebo mentions in his Ch. 15 four goals of the agricultural development policy in the Philippines: food security, competitiveness , alleviating poverty among farm families and protection of the environment. Of these goals alleviation of poverty is the most important, but the others require also the attention of the extension organisations. We think that the same goals are important in most Asian countries.

Most poor people live in families of small farmers and farm labourers (IFAD, 2001). Therefore it is a responsibility of the agricultural extension organisations to help to reduce their poverty. An important cause of their poverty is the decrease in the price of agricultural commodities in the world market (FAO, 2002: 12 and World Bank, 2005: 23) and in many local markets.

There are several possible ways to contribute to a reduction of the poverty among these farm families (see World Bank, 2005: 118 - 120). Often this requires finding ways to transfer money from the rich inside and outside the country to poor farmers by enabling these farmers to produces goods and services for which these rich feel a need.

1. *Improving their production techniques*: better control of plant diseases, use of fertilisers and irrigation water, choice of plant varieties, animal nutrition, etc. It has always been a major role of agricultural extension to contribute to these improvements and this should continue. However, we fear that this can not contribute enough to farm income to compensate the results of the further decrease in prices of agricultural products.
2. Part 2 of this book shows that *commercialisation of agriculture* can result in a considerable increase in income of some farm families, but not of all. This requires that these farmers have different

competencies than a subsistence farmer needs, such as:

- choosing high value products from which he can make a profit in the market,
- producing a better quality of these products at lower costs than his competitors inside and outside his country,
- selling these products for a good price; often this can only be done in co-operation with other farmers,
- finding information which is relevant for these tasks and applying this information in the decisions he makes.

It is clear that at present many Asian farmers do not have these competencies to such an extent that they are well able to compete with other farmers. Therefore, an important role of the extension service becomes to educate farmers to increase their managerial, entrepreneurial and decision making capabilities.

- Does the extension staff have the competence needed to provide this kind of education?
- How can they obtain the information needed to make their educational programme relevant for their target group?
- Are they well informed about the rapid changes in the market and in production technologies?

Which organisation is most suitable to develop an education programme on the production and marketing of high value products? Usually government officers have not much experience and training in marketing. In a pluralistic extension system other extension providers may better be able to provide this education.

3. With the present process of globalisation it becomes more and more important to help farmers to be

competitive in the world market. These markets are changing rapidly and also the available technologies are changing. Therefore farmers, extension agents and researchers should adjust the way in which they work at their changing environment. A researcher, who has earned a high status among his colleagues by doing the last 10 years good research on a certain problem, may not like to switch to studying a different problem, but that can be necessary for successful agricultural development.

4. So far a major goal of extension programmes was to help farmers to increase the yields per ha and per animal. However, the potential to *increase labour productivity* is much higher than the potential to increase land productivity (Ch. 1).
5. *Combining agricultural and non-agricultural sources of income* is a strategy which many farmers follow. They may also have a shop, weave cloth, work in the dry season in the city, etc. The market for touristic services is growing rapidly and this offers farmers, who are located in a beautiful area, e.g. in the mountains, income opportunities. Often there are more possibilities to increase their income from non-agricultural production than from agriculture. Therefore, scientists like Ellis (2000) recommend to explore these opportunities, but most extension organisations look only for opportunities inside agriculture. Are special extension programmes needed to teach farmers to earn money also from non-agricultural sources as was reported for Korea in Ch. 2? If so which organisation can conduct these programmes and how can their staff be trained for this new role?'
6. Many *farmers move out of agriculture or train their children for a non-farm job,* because they do not see

an opportunity to make a decent living on their farm. The discussion of labour productivity in Ch. 1 gives indications that some of these farmers may be right, but not all people who have moved out of agriculture are happy that they did. They may live in a difficult situation in an urban slum. There indications that the incidence of HIV/AIDS is relatively high among this group. It is clear that the decision whether or not to move out of agriculture is much more important for the welfare of the family than the decision how much and which fertilisers to give to their cereal crop. For most farmers, however, it is much easier to get advice on fertiliser use than on finding a job outside agriculture. Is that right? If not, who can give this advice and prepare the farm family for a successful live outside their village?

7. A single farmer faces many problems he can not solve, but it may be that they can be solved if he *works together with other farmers in a group*. That makes it possible to exert more power in the market, to have more influence on government policies and on extension and research programmes and to learn from the experiences and ideas of other group members. Some NGOs help farmers to establish these groups and to make them a success, but in most Asian countries farmers' organisations are still weak. Can and should extension organisations help to change this? If yes which organisations and how can they acquire the capability to do this well?

8. Agricultural extension mainly helps individuals to make better decisions. Many important problems, however, can only be solved through *collective decision making*. This is the case with decisions on natural resource management, environmental problems and increasing the power farmers have in the market. Supporting these decisions requires a

different extension approach as is discussed in several chapters.

9. Two key concepts in modern thinking agricultural extension, development and innovation are *partnership and learning*, as is illustrated in several chapters in this book and in the important book by Hall et al. (2004). They show that agricultural innovation does not only require changes at the farm level, but usually in the whole system including the supply of inputs and the marketing and processing of products. It also requires information from many different sources, including researchers in different disciplines and farmers.

Gao and Li show in Ch. 16 that learning from successful farmers is an important source of information for Chinese farmers. It could also be quite important for extension agents, but we are not sure that they make full use of this source in all countries and all extension organisations. Chs 5, 8, 12 and 13 show how extension agents and researchers can learn from them.

One task of extension agents is to facilitate the development of these partnerships and to stimulate that one learns from the experience of all relevant partners. Does this happen always? One of us said once in a lecture for extension scientists in an Asian country: "I see that in your country that coordination between animal and crop production takes place at the level of the farmer and at the level of the Prime Minister, but I do not see it anywhere in between". Everybody laughed, but unfortunately nobody said he was wrong. Our observations and several contributions to this book convinced us that improved partnerships between different actors are one of the most important ways to improve agricultural extension in Asia. Agriculture Man and Ecology (AME) helps NGOs in South India to realise these kinds of partnerships, but also elsewhere it may be useful to do this also. The Confederation of Indian Farmers' Associations insists that there should be a better coordination between the 21 Ministries dealing with different aspects of agriculture (The Hindu 11-12-2005).

10. A *relationship of trust between farmers and the persons, extension organisations, NGOs and commercial companies which support agricultural development* is quite important for the success of this.

A vegetable seed company may realise that farmers will only buy their seed in the long run, if this increases their income. If the profit from vegetable production with their seed increases with 100 $, because it gives a higher yield, better quality or lower costs of plant protection than the ordinary seed, the company may consider it fair that the farmer gets 90 $ and their company 10 $. So their seed will be much more expensive than the ordinary seed. Some traders will see a possibility to earn money quickly, because farmers can not see the difference between ordinary and high quality seed. So talking about this quality can be profitable. Farmers' associations can use the methodology of consumers unions to expose companies and traders who cheat farmers in this way. This may teach them that it is profitable to be honest.

NGOs are expected to work in the interests of poor farmers, but Hasannullah says in Ch. 7 that instead some NGOs try to increase the income of their leaders, even if this is at the expense of these farmers. It is also possible that NGOs like to serve the farmers' interest, but their staff has an urban background and a social science training and lacks the competence in agriculture to give advice which is profitable.

Commercial farmers should have the competence to choose partners they can trust.

11. Many agricultural extension organisations see as their *main role the transfer of information* from researchers working in government research institutes and universities to farmers. This is not enough because:
    - there are also other valuable sources of information such as researchers in commercial companies or outside the field of agriculture, farmers, consumers, partners in the market and policy makers,
    - there is no guarantee that farmers will use the information that is made available to them. They may also need help in the decision making process in which this information can be used to enhance their capability to make effective use of the available information in this process as is mentioned e.g. by Gao and Li and Avorn Opatpatanakit,
    - many extension agents tell farmers what they should do and how they should do this, but do not discuss with them why they should do this. If they did, they should increase the capability of their farmers to decide what they should do in a given situation. In this way they do not help to increase their managerial ability as much as possible.
12. Not all farmers are the same. They differ in education, in access to resources and markets and in their need to support from extension. That is a reason to provide this *support through a pluralistic extension system*. There is no reason why taxpayers should pay for the kind of agricultural extension for which input supply or marketing companies are

willing to pay and are well able to provide or for extension which specialised commercial farmers are willing to finance. However, much of the extension which poor farm families need can not be financed in this way. In order to alleviate their poverty this can only be paid by the government, but it may be provided by a farmers' association or an NGO.

13. There is pressure on extension organizations to support the process of commercialisation of agriculture, because this can increase income of some farmers, mainly the powerful farmers, and stimulate economic growth. However, this may not be the best solution for many underprivileged farmers, who are resource poor, live in a remote area and have a low level of education. At the same time their aspiration level increases as they become well aware that other families in their society increase their income. Part 3 of this book discusses possibilities for extension organizations to help also these people by discovering together with them, especially the women, solutions for their problems, which may work in their location and situation. Often this will not be through specialization, but by analysing their farm conditions as a whole and their risk bearing ability. They will not be able to pay for support from a privatised extension service, but in a pluralistic extension system it may be necessary to develop an extension organization, which specialises in support to poor farmers.

14. The *changes in* the role of agricultural extension, which are needed require different *competencies* from the *agricultural extension agents* as most of them do not have at this moment. People who are interested to become an agricultural extension officer should be able to follow an education programme which teaches these competencies. They should not

learn how to conduct an extension programme, but to analyse, which role of agricultural extension is required in their situation.

15. One may assume that the *average agricultural researcher is more intelligent than the average farmer*, but there is no doubt that some of the farmers and farm women are also very intelligent. In the situation in which they were raised it was not possible to get a good education, but lack of intelligence is not the reason that they did not earn a Ph.D. These farmers play an important rôle in agricultural development, e.g. by experimenting with new technologies and farming systems and as leaders of farmers' organisations. We think that their brains are the major resource for agricultural development in Asia, which is underutilised. Several chapters in this book show this can change by involving them in the design, implementation and evaluation of extension programmes.
16. Often there is a large *gap between the price farmers receive for their products and the price consumers pay* for the same product. If this gap could be reduced this would result in a considerable increase in farm income. Can agricultural research and extension contribute to closing this gap? If so, who can do this? Possibilities to decrease this gap are discussed by Dixie (2005).

Extension research

Now there are major changes needed in the role of extension, research on extension can and should make valuable contributions to decide which changes are desirable in a certain situation and how these changes can be realised. Let us give some examples of kinds of extension research which might be useful:

1. Anderson, Feder and Ganguly show in Ch. 6 that the large investments the World Bank has made in the T and V system often did not achieve the expected results. Many people, also at the World Bank, believe that it is profitable for a country to invest 1% of the value of its agricultural product in agricultural research. If they had also invested 1% of the T and V budget in research aiming at finding ways to improve the effectiveness of this extension system, it is very likely that this research would have increased the rate of return on the investment in the T and V system considerably on condition that this research was done by competent researchers from the relevant disciplines and not only by economists.
2. Hasannulah presents in Ch. 7 clear ideas how extension can be improved to make the commercialisation of agriculture more successful. We believe that many of his ideas are right, but he presents little evidence that this is true and we also do not know this evidence. Extension research may prevent that some of the money invested in implementing these ideas is misallocated.
3. This book shows differences of opinion on the optimal way to finance agricultural extension. Policy decisions on financing agricultural research are that important that we should not make these decisions without using research to learn from the experiences with the way of financing, but Anderson, Feder and Ganguly show that with the T & V system the quality of this research was not good enough to improve decision making.
4. Most of the extension research so far is on extension to transfer technologies by government financed agricultural extension programmes. The conclusions of this research may not be valid for extension programmes on marketing and on increasing the

managerial abilities of farmers or on extension programmes conducted by NGO's, farmers organizations and commercial companies. There is an urgent need for research which helps to learn from the experiences with these kinds of extension programmes.

5. A conclusion of this book is that for the alleviation of poverty it is important to start extension programmes on combining agricultural and non-agricultural sources of income and on the way members of farm families can make a living outside agriculture. Research is needed to make these programmes as successful as possible.
6. We said that partnerships between different stakeholders are important for realising successful agricultural development programmes, but ordering people to work in a partnership with others will not always result in a successful partnership. All actors should be convinced that it in their interest to cooperate with others, but it is no exception that they see each other as competitors. Hall et al. (2004) show how research can contribute to developing more successful partnerships. It may be profitable to apply their methodology also in e.g. the ATMA project, which is discussed in Ch. 8.
7. ICT has a large potential to make extension programmes more effective and to decrease their costs at the same time. To realise this potential it should be clear which role ICT is expected to play in an extension programme. Adoption research has shown e.g. that mass media can be very effective in increasing the awareness of farmers of and increase their interest in innovations, but these media are not effective in the decision making stage of the adoption process (Rogers, 1983). We do not yet know well what the role of ICT in this process can be.

Leeuwis with van den Ban (2004: 206) discuss some roles it can play in an extension programme:

- providing access to information, e.g. on the characteristics of the available rice varieties,
- securing feedback for learning, e.g. did a change in the feed ration given to cows increase their milk production,
- providing tailor made advice, e.g. how much and which feed to give to this cow which gives 14 litres milk a day,
- exchanging experiences among farmers.

Research is needed on how ICT can perform these and other roles effectively and on what is the optimal way to combine ICT with other communication channels in an extension programme.

In extension research it can be difficult to generalise from a pilot project where a new extension strategy has been tested, to the results this strategy will have when it is used on a large scale. For instance Ch. 8 shows rather convincingly that the ATMA project has resulted in an increase in the average income of the farmers in the 5% of Indian districts where this project was introduced. Because of this success it will now be introduced in another 50% of these districts. It is quite possible that this will result in an increase in the production of high value agricultural products. If this goes faster than the increase in demand, the prices of these products will decrease. Consumers should be happy when this happens, but farmers may think differently.

We say "should be" and not "will be", because many consumers may not realise that they profit from the improved agricultural extension strategy. That is one of the reasons why it is politically difficult to allocate a sufficient budget to the extension service.

References

Dixie, G. (2005). *Horticultural Marketing : Marketing Extension Guide 5. Rome, FAO.*

Ellis, F. (2000). *Rural livelihoods and diversity in developing countries.* Oxford University Press.

FAO (2002). *World agriculture: towards 2015 / 2030.* Rome, FAO.

Hall. A.J. et al. eds. (2004). *Innovations in innovation: reflections on partnership, institutions and learning.* Patancheru, ICRISAT.

IFAD (2001). *Rural poverty report 2001: The challenge of ending rural poverty.* Oxford University Press.

Leeuwis, C., with A.W. van den Ban (2004). *Communication for rural innovation: Rethinking agricultural extension.* Oxford, Blackwell Science.

Rogers, E. M. (1983). *Diffusion of innovations.* New York, Free Press, 3rd ed.

World Bank (2005). *Agricultural growth and the poor: An agenda for development.* Washington, D.C. World Bank.

Contributors

Abraham Blum (1928) has worked as orchard manager, agricultural teacher and curriculum developer, before he started the Extension Education Section at the Hebrew University of Jerusalem. There, he earned his B.Sc., M.SC. and Ph.D. degrees. Prof. Blum was the first incumbent of the Hayim Gvati chair in Agricultural Extension. His FAO book on Teaching and Learning in Agriculture was translated into Chinese. He has lectured at 50 universities and development centers in 24 countries, written 140 papers and development reports and produced five agricultural video films. Email : blum@agri.huji.ac.il.

Andrew J Hall is an Innovation Policy Researcher at the United Nations University (UNU) Institute for New Technology (INTECH), Maastricht, The Netherlands. In recent years he has worked on the development and application of the innovation systems concept as a way of exploring partnerships and institutional arrangement in agricultural research and development. UNU/INTECH, Keizer, Karel Plein 19, 6211 TC Maastricht, The

Netherlands. E-mail: hall@intech.unu.edu; andyhallmails @yahoo.com

Arun Balamatti started his professional career in NGO sector, currently works as the Executive Director, Agriculture, Man and Ecology Foundation, Bangalore, India. As a part of Doctoral Studies, he worked on ITKs and additionally obtained M.Sc in ARTES from Germany. Besides he has the professional training and exposure in Bangkok, Rome, and Ghana. His field of specialization is LEISA technologies for small dryland holders of the semi arid region. He can be contacted at AME Foundation, P.O.Box 7836, No.204, 100 ft. ring road, 3rd phase, Banashankari 2nd block, 3rd stage, Bangalore-560 085. E-mail- arun_balamatti@yahoo.com.

Ashok Jhunjhunwala is Professor of the Department of Electrical Engineering, Indian Institute of Technology, Chennai, India and was department Chair till recently. He received his B.Tech degree from IIT, Kanpur, and his MS and Ph.D degrees from the University of Maine. From 1979 to 1981, he was with Washington State University as Assistant Professor. Since 1981, he has been teaching at IIT, Madras.

Dr Jhunjhunwala leads the Telecommunications and Computer Networks group (TeNeT) at IIT Madras. This group is closely working with industry in the development of a number of Telecommunications and Computer Network Systems. Dr Jhunjhunwala is a Fellow of INAE, INSA and NAS and a member of Prime Minister Scientific Advisory Committee. He is a Director in the Board of SBI and also a Board member of several Telecom and IT companies in India, including Polaris, Sasken, Tejas and HTL, NRDC, and IDRBT. He can be contacted at : www.tenet.res.in.

Avorn Opatpatanakit (1960) works for Department of Agricultural Extension, Faculty of Agriculture, Chiang Mai University, Chiang Mai 50202 Thailand. She has also presently works as coordinator for the Thailand Research Fund (TRF). She has undertaken research projects and published academic articles on participatory agricultural extension, empowerment, community-based research and participatory action research. Email: avornopatpatanakit@yahoo.com.

Burton E. Swanson is Professor Emeritus of Rural Development at the University of Illinois. His career spans over 40 years, working in nearly 50 countries. The University recognized him with two distinguished faculty awards: Public Engagement (2002) and International Achievement (2004) and with numerous College awards. Dr. Swanson was founding president (1984) of the Association for International Agricultural and Extension Education and received its International Leadership Award (1991). Dr. Swanson has written or edited 8 books and more than 60 refereed or invited papers. Since 1990, he has worked extensively on World Bank extension projects in China and India; currently he is directing a USAID funded project in Egypt on Linking Farmers to Markets. Email: swansonb@uiuc.edu.

Cho Yeong-Cheol is Director of Extension Planning Division, Extension Service Bureau, Rural Development Administration, Korea. Dr Cho graduated from the College of Agriculture, Seoul National University (SNU) in Agricultural Extension Education with B.Sc. degree in 1978; obtained Masters' degree in Agriculture Extension Education during 1985 and Ph.D. in the same discipline in 1992 from SNU. Dr Cho is working as Extension Manager since 1981 in Rural Development Administration in Korea. Email: choyc@rda.go.kr.

Dwarakinath, R. born in a farmer's family in 1926 in Karnataka, India. He got his B.Sc., Ag (1949) from Mysore University, M.Sc. (1960) from University of Tennessee and Ph.D. (1973) from Cornell University, USA. His career was devoted to extension profession, from the field to the top, as Extension Keyman, Project Executive Officer, Director of Training (Extension), and Agriculture University, Director of Extension. Dr Dwarakinath also served as the State Director of Agriculture, and then as the Vice Chancellor, University of Agriculture Sciences, Karnataka. In 1980s, he was FAO Project Manager (Training Extension Functionaries) in Indonesia. Later, he was a consultant to Danida farm women projects in four Indian States. Recently, he was Chairman, Karnataka Agriculture Commission. Currently, he is the Chairman, AME Foundation devoted to promoting LEISA in the Southern states of India. Email : amebang@yahoo.co.in.

Elske van de Fliert is a communication specialist and ecologist with a doctorate in Communication and Innovation Studies from Wageningen Agricultural University, The Netherlands. She has worked for almost two decades on participatory research and learning in integrated pest and crop management, mainly in Asia, as researcher, lecturer and consultant for a range of international organisations. Her latest field assignment (2003-2006) was with the UN's Food and Agriculture Organisation's Regional Vegetable IPM Programme in Asia, where she managed the Vietnam and China components. As on July 2006, she will be an associate professor "Communication for Social Change", at the School of Journalism and

Communication, University of Queensland, Australia. Email : elskevandefliert@yahoo.com.

Gershon Feder is Senior Research Manager for Rural Development in the Development Research Group, World Bank since 1975. Dr Feder's extensive research work covers rural finance, technology diffusion, sovereign debt servicing capacity, land tenure and markets in developing countries and the impact of agricultural extension. Dr. Feder has a B.A. in Economics from Tel Aviv University, and a Ph.D. in Agricultural Economics from the University of California at Berkeley. His awards include the American Agricultural Economics Association's (AAEA) Award for Quality of Research and Publication of Enduring Quality. In 2005, he was inducted as a Fellow of the AAEA. He can be contacted at: gfeder@worldbank.org.

Hassanullah Mohammed, is a B.Ag., M.Sc. (Ag.) in Agronomy, M.Sc. in Extension Education and Ph.D. in Business Administration. Starting from 1961 he worked in several research, extension, teaching and specialist positions under government departments and corporations. From 1989 he became a freelance consultant and served FAO, UNDP, WB, CIDA, IFDC, USAID, DFID, UNOPS, IFAD, JICA. He published 19 scientific articles, 25 review articles, 14 popular articles, 9 books and booklets. Dr Hassanullah is a life member of the Institute of Management Consultants of Bangladesh, Association of International Agricultural and Extension Education, Indian Society of Extension Education. He is a life member and former President of Bangladesh Society of Agricultural Extension. Email : mhassan@citechco.net.

J.P. Singh earned his Bachelor Degree in Bio-Sciences from Agra University in 1973, Master's (1975) and

Doctoral (1986) Degrees in Agricultural Botany from Meerut University, India. Dr. Singh has diverse experience of working in Farm Management at G.B. Pant University, Pantnagar; research for improvement of Brassicas at IARI, New Delhi , formulation of strategies, implementation and monitoring of Centrally Sponsored Schemes of Technology Mission on Oilseeds, Pulses & Maize at national level . He also served as Secretary of National Oilseed & Vegetable Oil Development, Board a statutory Body under the Ministry of Agriculture. He also represented Indian Delegation to 15th Session of Codex Committee on Oil & Fat held in London, U.K. in November, 1996. Later part of his career, Dr. Singh worked for the Extension Reforms at national level. Recently on an assignment from the World Bank, he presented a paper on ATMA Model at the World Bank and USAID Headquarters in Washington D.C. in July, 2005. Email : jpsingh_doe@yahoo.co.in.

Jock R. Anderson, farmer, Monto, Queensland, studied agricultural science at the University of Queensland, and agricultural economics at the University of New England, Armidale, where he was Professor of Agricultural Economics, and Dean of the Faculty of Economic Studies, with off-campus engagements such as Chief Economist of ABARE, and Director of the 1985 Impact Study of the CGIAR. Post 1988 in the World Bank he held several positions including Adviser, Strategy and Policy, retiring to Washington-based consulting in 2003. He is a Fellow of AAEA and the Academy of the Social Sciences in Australia, among others. Email : janderson@worldbank.org.

Kalim Qamar, M. earned M.Sc. from Pakistan, M.S. from American University of Beirut, and Ph.D. from

Cornell University, USA. Later, he taught rural non-formal education at University of Massachusetts, USA. Dr. Qamar served as Advisor on Rural Development Training in Papua New Guinea, and as Team Leader/extension, communication and training specialist on World Bank/UNDP-funded projects in Indonesia, Nepal and Iraq. At the headquarters of Food and Agriculture Organization of the United Nations (FAO) in Rome, Dr. Qamar first served as Country Projects Officer, and is now Senior Officer (Agricultural Training & Extension). His international experience covers over 50 countries of Asia and the Pacific, Middle- and Near-East, Africa, Caribbean, Europe, and North America. E-mail: Kalim.Qamar@fao.org.

Nagaratna Biradar (1970), completed Ph.D. doctoral studies from University of Agricultural Sciences, Dharwad, India and P.G. Diploma in Agricultural Extension management from MANAGE, Hyderabad. Ms. Biradar was trained in PRA at Aga Khan Foundation, Ahmedabad and Development Oriented Research in Agriculture from ICRA, Wageningen. She is involved in many interdisciplinary projects funded by ICAR and other foreign donors. She also acts as facilitator for various NGOs running projects funded by Ministry of Rural Development, Government of India, and is recognized as PG teacher of UAS, Dharwad. Currently she is working as Sr. Scientist, Agricultural Extension at IGFRI-Regional station, UAS Campus, Dharwad-580 005, Karnataka, India. E-mail:nagaratna123@rediffmail.com.

Nguyen Van Linh (1956) graduated from the National Economics University in Hanoi (Vietnam) in 1978, completed the MSc programme on Management of Agricultural Knowledge System and got the PhD degree in Social Sciences from Wageningen

University, The Netherlands. He worked for Vietnam Agricultural Science Institute for 24 years as a Researcher of Farming System Research department. He experienced on working with many agriculture and rural development projects in Vietnam. He can be reached at nvlinh@fpt.vn.

Qijie Gao (1963) studied at Hohenheim University in Germany and the University of Sydney in Australia, was the first in China to obtain Ph.D. degree in Rural Development and Extension. He formerly worked as Director of Department of Rural Development and Extension at China Agricultural University (CAU), is now Director of Centre for Extension and Innovation Management at CAU, and Deputy Director of China High & New Agrotechnology Society. He has worked as Director of many projects and published many books and papers on Rural Extension, Development and Innovation. His address : College of Development, China Agricultural University, Beijing 1000094, P.R. China, Email: gaoqj@cau.edu.cn.

Rangnekar, D.V. (1938) born at Mumbai, India. Graduated in Veterinary Science and Animal Husbandry from Veterinary College at Mhow, M.P., Masters and Ph.D. in Animal Nutrition from Indian Veterinary Research Institute at Izatnagar in 1964 and started his Professional career at Amul Dairy, Anand and then at Cyanamid India Ltd. Dr Rangnekar joined Bharatiya Agro-Industries Foundation (BAIF) at Urulikanehau, Pune in 1968 as the first member of its development team and worked for 31 years and retired as its Senior Vice President. During 2000-2004, Dr Rangnekar was associated with the National Dairy Development Board (NDDB), Anand as dairy production consultant and presently working with National

Innovation Foundation, Ahmedabad, India. Email : dattavr@rediffmail.com.

Rasheed Sulaiman, V. is a Senior Scientist at National Centre for Agricultural Economics and Policy Research (NCAP), New Delhi, India. Has been working on policy aspects of agricultural technology generation and diffusion at NCAP since 1993. In recent years he has studied institutional innovation in extension, including privatization models as well as exploring extension as part of broadbased development interventions in the agricultural sector. In this work he has increasingly used the innovation systems concept as a way of thinking about technology transfer in a more holistic sense. NCAP, PO Box 11305, Library Avenue, Pusa, New Delhi 110 012, India, E-mail: rasheed.sulaiman@gmail.com.

Samanta, R.K. (1950), Ph.D. (Agricultural Extension), worked as Director, Agricultural Extension and Communication at National Institute of Agricultural Extension Management (MANAGE), Government of India, Hyderabad and has published some important books on agricultural extension, its management and rural development. Some prominent of them are : Agricultural Extension in Changing World Perspective (1991); Training Methods for Management and Development (1993); Management of Agricultural Extension in the Global Perspective (1997); KVK (Farm Science Centre) – the Capacity Builder of Farmers (2002); and Empowering Rural Women (2005). Dr Samanta's professional training has been in India and USA with Cornell, Syracuse and the Ohio State Universities. He has also been IAEA consultant twice in Bangladesh during 1996 and 1998 and presently Joint Director at National Academy of Agricultural Research Management (NAARM), Hyderabad 500 030, India, Email:

ranajitsamanta123@yahoo.com / samanta@naarm.ernet.in.

Samuel T. Mancebo, a Filipino born in 1941 has a Ph.D. in Agricultural Education in 1979 at the University of the Philippines Los Banos and had his post doctoral studies at the Ohio State University in 1982 as a World Bank Fellow. He is a Professor in Agricultural Education, a former Dean of the UPLB College of Public Affairs, a Project Leader/Director of various researches and consultant in Agriculture Education, Extension, Curriculum Development, Institutional Development and Management to various organizations such as the World Bank, Asian Development Bank, European Union, Aus-Aid, FAO, UNESCO and the World Agroforestry Centre working in many countries in Asia. He has authored more than 65 articles, manuals, books and research papers and reports to his credit. Email: gsam8941@yahoo.com.

Sangamitra Ramachander is Research Manager at n-Logue Communications Private Limited, Chennai, India. n-Logue is a Rural Service Provider, with a network of over 2000 rural Internet kiosks spread across India. Her current research focuses on gaining insights into the strengths and sustainability of the rural kiosk business model, through operator and user perspectives. She is an Associate of the TeNeT Group, which incubated n-Logue, and works with a team that builds kiosk services. Prior to joining n-Logue, she has researched in the area of food security and social safety nets at the International Food Policy Research Institute (Washington DC) and climate change impacts at the Joint Global Change Research Institute (College Park, MD). Sangamitra has two Master's degrees in Economics, from the University of Maryland, College Park (2003) and the

Madras School of Economics, Chennai (2001). A native speaker of Tamil, she holds a diploma in French. She can be reached at: sangamitra.r@tenet.res.in

Singh, K.M. earned his Master's (1983) and Doctoral (1986) degrees in Agricultural Economics from G.B. Pant University, Pantnager, and Banaras Hindu University, Varanasi, respectively. He joined the Rajendra Agricultural University, Bihar in 1985 as Assistant Professor and became Professor of Agricultural Economics in 1998. Dr. Singh has authored several books, including two text books on Agricultural Marketing and International Trade, along with over fifty research articles. Recently he presented a paper at World Bank and USAID headquarters in Washington DC on Developing Supply Chains for Medicinal Plants. He presently works as Project Director of ATMA, Patna. Email: krishna_singh2000in@yahoo.com.

Sridhar, K. specialized in Breeding of forage crops completed doctoral studies from IARI, Pusa. Associated with Indo-UK forage production project, has been active in participatory technology development and diffusion process of IGFRI. Currently working on forage sorghum, has published several research papers, besides recognised as resource person for state and central government training programs. He can be contacted at IGFRI-Regional station, UAS Campus, Dharwad, Karnataka, India. E-mail- s_kandlakunta @rediffmail.com.

Sushma Ganguly currently holds the position of Sector Manager, Agriculture and Rural Development Department, in the Environmentally and Socially Sustainable Development Vice Presidency at World Bank. Ms. Ganguly, an Indian national, joined the

Bank as a Young Professional in May 1980, working in the Europe, Middle East and North Africa (EMENA) Projects Department ("old" Yugoslavia) and in the Latin America and Caribbean Programs Department (Colombia). Beginning in 1981 she worked as an Economist in agriculture divisions in EMENA (Turkey & Yugoslavia) and in the East Asia and Pacific Region (Indonesia). In 1987 she was promoted to Senior Economist and in 1991 to Principal Economist. Two years later Ms. Ganguly became Division Chief in the Agriculture and Environment Division of Eastern Africa Department. In 1996, she was appointed Technical Manager, and then Sector Manager, Rural Development Operations, Eastern and Southern Africa, covering almost the entire Anglophone Africa.

Van den Ban, A.W. (1928), earned his M.Sc. and Ph.D. degrees at Wageningen Agricultural University in the Netherlands. At this University he started the Department of Extension Education. He worked as a consultant on agricultural extension in 25 countries, including 13 Asian countries. Of his textbook on agricultural extension some 70000 copies have been printed in 13 languages. He can be reached at Foulkesweg 82A, 6703 BS Wageningen, The Netherlands. Email: avandenban@planet.nl.

Xiaoyun Li (1961) graduated from Ningxia Agricultural College and earned Ph.D. degree at China Agricultural University (CAU). He visited many countries and made advanced studies in Germany and the Netherlands. He formerly worked as Director of Centre for Integrated Agricultural Develoment at CAU, is now the Director of College of Humanities and Development at CAU. He has worked as Director of many products and published many books and papers especially on rural development, training, rural poverty alleviation and Gender. He can be reached at College of Development, China Agricultural University, Beijing 100094, P.R. China, Email: xiaoyun@cau.edu.cn.

Yang Seung – Choon is working as Senior Extension Specialist in Extension Planning Division in Rural Development Administration, Korea. Dr Yang graduated from the College of Agriculture, Seoul National University (SNU) in Agricultural Extension Education with B.Sc. degree in 1978; obtained Masters' degree in Agriculture Extension Education in 1984 and Ph.D. in the same discipline in 1994 from SNU. Dr Yang is working as Extension Specialist since 1985 in Rural Development Administration in Korea. Email : ysc@rda.go.kr.